THE GENUS
PLEIONE

A KEW MAGAZINE MONOGRAPH

The Genus PLEIONE

Phillip Cribb and Ian Butterfield
assisted by Tang Chen-zi (C. Z. Tang)

Illustrations by
Christabel King
and
Rodella Purves and Margaret Stones

Series Editors
Christopher Grey-Wilson
Victoria Matthews

The Royal Botanic Gardens, Kew
in association with Christopher Helm
and Timber Press

Christopher Helm (Publishers) Ltd, Imperial House,
21–25 North Street, Bromley, Kent BR1 1SD

ISBN 0-7470-0422-6

A CIP catalogue record for this book is available from
the British Library

First published in North America in 1988 by
Timber Press
9999 SW Wilshire
Portland
Oregon 97225
USA

Library of Congress Cataloging-in-Publication Data

Cribb, Phillip.
The genus Pleione / Phillip Cribb and Ian Butterfield assisted by
Tang Chen-zi (C.Z. Tang); illustrations by Christabel King and
Rodella Purves and Margaret Stones.
p. cm.
Rev. and expanded ed. of an article with the same title first
appearing in Curtis's botanical magazine, v. 184, pt. 3 (1983).
A Kew magazine monograph.
Bibliography: p.
Includes index.
ISBN 0-88192-126-2
1. Pleione. 2. Pleione—Classification. I. Butterfield, Ian.
II. Tang, Chen-zi. III. Title.
QK495.064C75 1988
584'.15—dc19 88-16930
CIP

Photoset by Paston Press, Loddon, Norfolk
Printed and bound in Great Britain by Butler and Tanner, Frome, Somerset

To Lisa Sainsbury

CONTENTS

LIST OF COLOUR PLATES

ACKNOWLEDGEMENTS

The authors would like to thank Drs D. Harberd, H. Pfennig, R. Stergianou and C. Grey-Wilson and Messrs C. Bailes, N. Trudel and H. Pinkepank for helpful advice on the text and for the loan of colour photographs.

A revision of the genus *Pleione* by Cribb, Butterfield and Tang first appeared in *Curtis's Botanical Magazine* in 1983 (Vol. 184, part 3). This work has been updated in the light of more recent research and the text expanded for this new book.

We would particularly like to thank Christopher Grey-Wilson and Victoria Matthews, the Series Editors, for their patience and guidance in seeing this work through to publication.

INTRODUCTION

The genus *Pleione* is a group of small, semi-hardy, popular orchids of great beauty. They are on the whole easy plants to cultivate, indeed many can be grown in the average home, given the right conditions. For this reason they have been popularly called 'Window-sill Orchids'. In recent years artificial hybridization between various species and cultivated forms has led to a wide variety of exciting and exotic-looking hybrids that can be purchased, unlike many orchids, at a fairly reasonable price. Both professional and amateur growers find them equally appealing. It is not therefore surprising that in recent years there has been a remarkable resurgence of interest in the genus. This has been stimulated in part by the introduction of fresh material from the wild, especially from China; *Pleione scopulorum* and possibly *P. grandiflora* have lately been introduced into cultivation from China as well as the recently described *P. coronaria* from central Nepal. Furthermore, new clones of *P. bulbocodioides*, *P. forrestii*, *P. speciosa* and *P. yunnanensis* have also appeared in cultivation in the past three or four years.

The three authors of this work have had a good deal of experience with the subject. Dr Phillip Cribb is a taxonomic botanist at the Royal Botanic Gardens, Kew, specializing in the Orchidaceae. Mr Ian Butterfield runs a nursery specializing in the growing of pleiones and dahlias and has done a good deal of hybridization work in the genus *Pleione*. Mr C. Tang is a horticulturist and botanist at the South China Botanic Garden, in Canton.

Horticulturists and gardeners may be alarmed by name changes for well-known species. This has been unavoidable due to earlier confusion and subsequent misidentification. Thus the plants in cultivation as both *P. delavayi* (2*x*) and *P. yunnanensis* (6*x*) are correctly *P. bulbocodioides*; *P. forrestii* of cultivation is now known to be of hybrid origin and should be called *P.* × *confusa*; *P. pogonioides* should be known as *P. speciosa*. The true *P. forrestii* and *P. yunnanensis*, as has been noted above, have been recently collected in the wild in China so these two charming species are now available for comparison.

Christopher Grey-Wilson
Victoria Matthews

HISTORY OF THE GENUS

The genus *Pleione* was established in 1825 by David Don based on two Himalayan species, *P. praecox* and *P. humilis*, both previously figured and described by Sir James Smith as species of *Epidendrum*. Subsequent authors, notably John Lindley (1830 & 1854), H. G. Reichenbach (1864) and Sir Joseph Dalton Hooker (1890), did not follow Don, but instead reduced *Pleione* to a section of the large genus *Coelogyne*. As Lindley (1854) stated—'There is something so peculiar in the plants called *Pleione* by Don, that it would be desirable to find some means of separating them from *Coelogyne* . . . But . . . after all I find it necessary to leave them as an alpine form of *Coelogyne* in the hope that future observers will discover some sound generic characters.'

It was left to Rolfe (1903) to resurrect the genus. Whilst recognizing the floral similarities of *Pleione* and *Coelogyne* he nevertheless distinguished the former on the grounds that 'the lip is quite different in shape and details; added to which there is the very differently shaped annual pseudobulbs, the thin deciduous leaves and the peculiar inflorescence'.

Subsequent authors such as Pfitzer & Kraenzlin (1907), Schlechter (1914, 1919 & 1922) and Hunt & Vosa (1971) have all followed Rolfe in accepting *Pleione*.

THE MORPHOLOGY OF PLEIONE

Pleiones are very simple plants in comparison with most orchids. Their habit is dominated by their annual life-cycle with an active building phase through the spring and early summer, and a dormant phase in the winter. The annual nature of the growth of *Pleione* has produced morphological adaptations that distinguish the genus from its close ally *Coelogyne*. Coelogynes are plants of warmer tropical and subtropical climates and lack a dormant period.

THE PSEUDOBULBS AND LEAVES

The pseudobulbs of *Pleione* comprise a single node and develop from the base of the previous year's growth. The new shoot produced in the spring swells as the plant begins to grow, eventually bursting its enclosing sheaths as they dry. In all of the spring-flowering species the pseudobulbs are smooth-surfaced. In species related to *P. formosana* they are usually ovoid or dorso-ventrally compressed-ovoid in shape and can develop to a few centimetres in diameter in well-grown plants. The pseudobulbs of *P. humilis, P. albiflora, P. coronaria, P. forrestii* and *P. grandiflora* are often conical in shape, and, in the first two, a conspicuous long slender neck is often present. In these two, and in *P. coronaria*, the pseudobulbs have at the apex a conspicuous brittle, brown crown of tissue left by the deciduous leaf when it falls. In *P. hookeriana* and *P. scopulorum*, the pseudobulbs are usually much smaller and often scarcely one centimetre in diameter. The small size of these pseudobulbs may be one reason for the relative difficulty experienced in growing these species successfully, particularly as they tend to desiccate rather readily in dormancy if kept in too dry an atmosphere.

Most of the spring-flowering species have dark green pseudobulbs but various degrees of purple-flushing are found in some clones. This is particularly common in *P. bulbocodioides* and in *P. formosana*. In the wild, several species, such as *P. yunnanensis*, grow with their pseudobulbs buried a centimetre or more deep in leaf-litter, moss or friable soil. When removed these pseudobulbs are usually white but, when grown in cultivation buried up to their necks, the pseudobulbs turn green.

The autumn-flowering species, *P. praecox* and *P. maculata*, together with *P.* × *lagenaria*, have the most distinctive pseudobulbs in the genus. These are flask- or bottle-shaped rather than ovoid or conical. The sheathing bracts around the developing shoots split and fray as the new pseudobulbs develop and a net-like reticulum of dried-up veins is left to sheath the maturing pseudobulbs. The pseudobulbs of *P. praecox* are also delightfully mottled with red and green, and warty, while those of *P. maculata* are green or brownish but lack small warts.

Bulbils are produced at the apex of the pseudobulbs of many species. This is particularly common in *P. humilis* and *P. hookeriana* where the mature pseudobulbs can often sport a medusa-like head of bulbils at their tips. *Pleione formosana* and its allies also produce bulbils but never so many. There is evidence to suggest that in some species these appear more frequently in cultivation than in the wild, where the pseudobulbs are often buried below the surface. Bulbils are a means of rapid reproduction for those species that produce them and lead to the formation of colonies of genetically identical plants.

When growth begins in the spring, new roots are produced from the base of the new shoot at the side of the mature pseudobulb. The new roots are slender and elongate rapidly in their search for moisture. In the wild, root growth often begins before the rains with the atmospheric humidity and dew providing enough moisture to stimulate root growth. In cultivation watering should be carefully controlled at first to prevent root loss from rot.

The species of *Pleione* divide neatly not only on their flowering time but also on their leaf number. The autumn-flowering species and *P. scopulorum* have two-leaved pseudobulbs while those of all of the other species have one leaf only. The leaves are plicate in all species and range in shape from lanceolate to oblanceolate with a drawn-out tip. Those of *P. yunnanensis* are often rather narrow but otherwise the species are difficult to distinguish on leaf characters alone. The leaf, or leaves, develop from the new shoot apex, emerging from the sheaths that cover the shoot. The sheaths of the autumn-flowering species are quite distinctive and unlike those of the other species; in *P. maculata*, the sheaths are green and inflated while, in *P. praecox*, they are mottled with red and green like the pseudobulbs, and warty.

THE INFLORESCENCE

In all pleiones the inflorescence emerges from the apex of the new shoot, either at the beginning of the growing season in the spring-flowering species or at the end in the autumn-flowering ones. In the former, it develops as the new leaf emerges and as the new roots begin to grow. In contrast, the flowers appear as the old leaves fall in the autumn-flowering species, but again it is from the new shoot. The normal condition in most pleiones is that the inflorescence bears a single flower but in cultivation at least two flowers per stem are not infrequent. Well-grown plants often produce two or even three new shoots from an old pseudobulb and each of these can flower and eventually produce a new pseudobulb. The new shoot usually develops in a slight depression on the edge of the mature pseudobulb.

THE FLOWER

The flowers are quite typical of the tribe *Coelogyneae* in structure and are subtended by a bract which is green or rose-pink and usually as long as the ovary. The ovary, as in all orchids, is inferior and is marked with six longitudinal grooves; in the mature fruit dehiscence occurs along three of these. At the apex of the ovary the floral segments are borne in three whorls.

The outer whorl comprises three petaloid sepals which are free to the base, similar in shape and usually the same colour as the petals. The next whorl has two showy lateral petals and a third much-modified one—the lip or labellum. The lip is larger than the petals, of a distinctive shape, often three-lobed, and usually bears a callus on its upper surface. The callus features are amongst the most useful for identifying the species. In *P. hookeriana*, *P. humilis* and *P. albiflora*, for example, the callus comprises several longitudinal lines of long hairs; in *P. praecox* and *P. maculata* the callus comprises lines of papillae; in *P. yunnanensis*, *P. coronaria* and *P. forrestii* the callus-ridges are entire thin lamellae of tissue running from the base almost to the apex of the lip. The number of callus-ridges, their degree of laceration and their marginal undulation are all useful taxonomic characters. This is scarcely surprising because the callus-ridges are guides that conduct the pollinator so that it is exactly positioned to pick up the pollinia or to place the pollinia in the stigmatic cavity of the flower. Pollination is almost certainly species-specific in *Pleione* and, therefore, callus

Plate 1

Pleione maculata (top)

Pleione × *lagenaria* (bottom) CHRISTABEL KING

Plate 2

Pleione praecox CHRISTABEL KING

structure is similarly species-specific. Although no study has been made of the pollinators of *Pleione*, they are almost certainly species of bumble bees.

In a few species, such as *P. albiflora* and *P. forrestii*, the lip has a distinctly saccate base which is almost hidden by the base of the lateral sepals.

The lip is also the most brightly coloured organ in the flower, again linked to its specific role in attracting pollinators. In most species the lip is variously marked with purple spots or dashes while the callus and the area around the callus are also distinctively coloured, usually in yellow.

The third floral whorl comprises the sexual organs which are united into a single organ, the column. The column is a distinctive feature of all orchids and is the result of the reduction in number and fusion of the male and female sexual parts. The column-stalk results from the fusion of the style and stamen-filaments. In *Pleione*, the male organs are reduced to a single fertile anther, which lies at the apex of the column and consists of an anther-cap attached by a very short stalk and enclosing four pollen masses or pollinia. The pollinia are obovate in shape and bilaterally flattened and each contains thousands of pollen grains in tetrads; these are united into discrete and rather hard waxy masses (each called a pollinium). This adaptation is linked to the pollination syndrome, for the pollinator picks up the pollinia by means of a sticky mass (viscidium) which sticks to the pollinator (often to its head or back), which then transfers them to the stigmatic cavity of another flower, thus ensuring cross-pollination.

The stigmatic cavity in *Pleione* is placed towards the apex of the column but on its lower side just behind the anther. The cavity contains a sticky secretion and there is a ridge-like rostellum in front of the cavity. These enable pollinia to be removed easily from the back of the pollinator when it leaves the flower.

MONSTROUS FLOWERS

Monstrous flowers are not at all common in *Pleione*. We have seen a form of *P. formosana* which produced four petals and a column, but no lip. One plant of *P. formosana* 'Oriental Splendour' produced a flower with seven petals but the following year had normal flowers.

Some clones of *P. forrestii* have produced malformed flowers with a reduced number of floral segments and in another the petals were joined to the lip, but these all produced normal flowers when grown under better conditions. *P.* Novarupta 'Goshawk' also had its petals joined to the lip when it first flowered but later produced normal flowers.

We have once seen a *P.* Eiger which produced two lips in a flower, one on top of the other. Likewise, a clone of *P. formosana* produced two lips in a flower but these lay side by side.

Our experience suggests that under poor culture conditions pleiones can produce occasional malformed flowers but that good cultivation will usually cure these problems.

THE FRUIT-CAPSULE AND SEEDS

Successful cross-fertilization leads to the development of the ovary into a mature capsular fruit which is fusiform and longitudinally ridged. When ripe this dehisces along three lines releasing thousands of dust-like seeds which are then wind-dispersed. Subsequent germination can occur only in the presence of the right mycorrhizal fungus because the *Pleione* seed carries no endosperm. The embryo relies for its food supply upon the mycorrhizal fungus which penetrates the embryo through one end, invades the inner cells but is then itself attacked by the orchid cells which remove sugars and micronutrients from the fungal cells. In the laboratory, *Pleione* seeds can be induced to germinate and grow *in vitro* in the absence of the fungus but only if free sugars and micronutrients are supplied in an agar gel.

An unpublished study of seed morphology in *Pleione* (Cribb & Page, in prep.) indicates that the *Pleione* species so far studied fall into three groups based on seed structure and ornamentation. Both *P. praecox* and *P. maculata* have short fusiform seeds with a pattern of narrowly oblong lacunae surrounded by a slightly undulate, raised, double-banded marginal rim. In *P. yunnanensis* and *P. forrestii* the seeds are smaller and still fusiform, but the pattern has altered with more regularly oblong, wider lacunae surrounded by a raised, apparently single-banded rim. Finally, in *P. scopulorum* and *P. hookeriana*, the seeds are much longer and cylindrical, closely resembling those of the *Coelogyne* and *Bletilla* seeds studied. In these the lacunae are obliquely oblong and again surrounded by an apparently single-banded rim.

No seed has yet been obtained of *P. coronaria, P. humilis, P. albiflora* or any of the 'bulbocodioides complex' and until these are available it would seem premature to state any more definite conclusions.

CYTOLOGY

Our knowledge of chromosome number in *Pleione* has been summarized by Hunt & Vosa (1971), Wimber & Cribb (1981) and Stergianou (1987). The basic chromosome number in the genus is x = 20 and most of the species have proved to be diploid. The only exceptions have been found in species of the taxonomically difficult 'bulbocodioides complex' where diploid, tetraploid and hexaploid counts have been recorded and it may well be that polyploidy is responsible for at least some of the difficulties in interpreting this group.

Cytological studies have also confirmed the hybrid origin of *P.* × *confusa* (syn. *P. forrestii* cult.), see Wimber & Cribb (1981).

A large number of counts have also been made on artifically raised hybrids. A high proportion of these have been shown to be either polyploids or aneuploids (the latter having fewer or more chromosomes than an exact multiple of the haploid number), and this is often reflected in the fertility of these clones in further hybridization programmes.

CULTIVATION

Pleiones grow in the wild in a wide variety of habitats at altitudes from about 1000 m up to 4200 m. They can be found growing on almost bare branches of forest trees, in moss on the branches and trunks of trees, or on fallen logs, on moss-covered rocks, in moss- and soil-filled pockets on cliff faces, in leaf-litter under shrubs and small trees and even sometimes in grassland. All situations have one feature in common—free drainage.

During the growing season the climate is wet and relatively warm, but during the winter months, for most species, it is cold and dry. The plants are completely dormant during the winter with no leaves or live roots. Those growing at higher altitudes may be covered by snow for part of the time.

GENERAL RULES

There are three essential requirements for the successful cultivation of pleiones:

1. Do not overwater during the spring when the new roots are emerging at the base of the new shoots. The compost must be kept only slightly damp to force the new roots to search for moisture.
2. When the plants have developed a good root system and are growing strongly they should be fed regularly.
3. Plants should be given a suitable rest period during the winter months with low temperatures, 0–2°C, for at least part of the time.

Pleiones are shallow-rooted and grow best in pans or half-pots. If plastic pots are used, then extra holes must be drilled in them to ensure free drainage. A 25 cm diameter pan requires about twenty 6 mm holes in the base. Providing that there are plenty of drainage holes, pots will not require crocks below the compost. It is wise to sterilize all pans by scrubbing them well before use in Jeyes fluid, or a similar agent, in order to reduce the possibility of fungal infections later.

COMPOST

Many combinations of compost have been successfully used for growing pleiones. One which can be particularly recommended is, by volume, 6 parts ungraded orchid bark with the dust and fine particles (up to 5 mm) sieved out, 1 part coarse grade perlite, 1 part sphagnum moss chopped into 6 mm lengths and 1 part sphagnum moss peat which has been thoroughly broken up. This coarse compost is used in the bottom two-thirds of the pan. A similar, but less coarse, compost is used above, using finer grade bark and sphagnum chopped into shorter lengths, and this is ideal for packing around newly planted pseudobulbs.

Another compost that produces excellent growth is 6 parts medium grade orchid bark, 1 part coarse perlite, 1 part medium charcoal, 1 part beech leaves, 1 part oak leaves and 3 parts chopped sphagnum moss. The beech and oak leaves are collected when freshly fallen in the autumn and are chopped up in a garden shredder. No finer grade compost is used for top dressing the newly planted pseudobulbs in this mixture.

Fibrous loam can be used instead of bark in the compost, but garden soil is most unsuitable. If loam is used then it is best to add a generous portion of coarse grit, or extra perlite, to the mixture. Some growers add small proportions of half-decayed oak or beech leaves to their compost, or even dried horse or cow dung that has been rubbed through a coarse sieve. The chopped up fronds and stems of bracken, *Pteridium aquilinum*, are also a useful addition for composts that tend to stay too wet.

All composts can be kept 'sweet' by adding a small proportion of fine grade charcoal.

A GROWER'S CALENDAR

Pleione pseudobulbs are usually bought during December or January when they are completely dormant and that is a logical place to start this calendar.

January–February. Pseudobulbs should be potted up, preferably before they start into growth. Pans are filled to within about 3 cm of the rim with coarse compost which is then dusted with bone-meal.

The pseudobulbs are then placed on the compost allowing only 1 cm between each (for large pseudobulbs). This may seem rather crowded but plants generally do better when closely set, this in effect simulating the way many grow in nature. Fine grade compost is then poured around the pseudobulbs until they are almost covered. If a loam-based compost is used then it is wiser to insert only the basal quarter of the pseudobulbs in the compost.

Ideally the pans should then be placed in a cold greenhouse or frame with a minimum night temperature of 0–10°C. No water should be given until the flower-buds can be seen.

March. New growths and/or flower-buds will appear towards the end of the month as the days lengthen and the temperatures rise. A little water can then be given, but only enough just to moisten the compost and encourage new roots to form. It is important to remember that *Pleione* roots never branch, unlike those of most orchids, and damaged roots will die back. Root damage is probably the main way in which plants are lost at this time of year. If the pans are inadvertently overwatered they should be allowed practically to dry out before more careful watering is resumed.

Whilst in active growth plants seem to benefit greatly from a good movement of air, so the installation of a circulatory fan in the greenhouse will prove beneficial, indeed some of the high-altitude species grow best when placed very close to a fan. Alternatively plants should be placed close to a ventilator.

At this time of the year the young leaves are tender and must be shaded from strong sunshine to prevent scorching.

April–May. As soon as the flowers begin to fade the leaves commence a period of rapid growth. This is the time to start liquid feeding. Any liquid fertilizer can be used, but a well-balanced one of nitrogen, phosphate and potassium (NPK) in the ratio 3:2:2, plus trace elements, gives the best results if given at half the normal recommended dilution rate.

June–August. During this period plants benefit from plenty of moisture in the atmosphere and also in the pans. However, care must still be taken to ensure very free drainage, as the plant would enjoy in the wild. They respond in particular to having their leaves sprayed all over during the evening, especially after a hot sunny day. The floor of the greenhouse should also be damped down to create a moist atmosphere overnight.

The day temperatures at this time of the year are best below about 32°C, but if they do rise higher then the plants will not come to too much harm provided the atmosphere is kept moist. Gardeners often overcome such problems by placing the plants out-of-doors in a cool semi-shaded corner of the garden during this period.

In the middle of July it is advisable to change the liquid feed to one with a higher proportion of phosphate and potash (NPK 2:3:3 is suggested). This will encourage the pseudobulbs to ripen and ensure good flower-bud formation for the following season. Feeding should continue every week until the leaves begin to turn yellow.

September onwards. The leaves of *P. speciosa* and *P. limprichtii* are the first to turn yellow, generally in late September. Most of the other species and cultivars soon follow, the leaves finally turning brown and dropping. Growth has by then more or less ceased for the season and the pseudobulbs enter the winter rest period. About three weeks after leaf-fall, the roots die off so it is best to stop watering when the leaves fall. Some plants may not become completely dormant until well into December.

This is also the time of year that the autumn-flowering species burst into bloom. Despite their late-flowering habit even these plants are entering their dormant period during late autumn.

The pseudobulbs of spring-flowering species and hybrids can be cleaned once they are completely dormant, in readiness for the repotting which is best undertaken every year. Cleaning consists of removing the old shrivelled pseudobulbs and trimming back the dead roots to a tuft about 6 mm long. The root-tuft is important in supporting the replanted pseudobulbs until the new roots grow the following spring.

All the large flowering-sized pseudobulbs should be repotted together so that a more uniform pan of flowers is produced. The smaller ones, and bulbils produced in abundance by some species and cultivars, should be kept together to be grown on to flowering size in a season or two.

Additional cultivation details applicable to individual species can be found under each species in the main text.

OUTDOOR CULTIVATION

Pleiones can be successfully grown outdoors, but great care must be taken if results comparable with glasshouse cultivation are to be achieved. The site must be carefully selected as too much moisture, combined with frost during the winter, will almost certainly kill them. The drainage must be excellent—either by planting on a slope, growing in a raised bed containing plenty of grit or by planting in the vertical cracks of a peat wall. It is also advisable to protect the pseudobulbs against rain during the winter, either by placing a sheet of glass over them, or by planting them under shrubs such as camellias or rhododendrons.

The chosen site in the garden should not receive early morning sun while frost is still on the ground as the plants cannot tolerate being frozen and thawed regularly. The compost for planting should contain plenty of humus either in the form of leaf-mould or peat and also coarse grit and/or coarse perlite. Species such as *P. limprichtii* will thrive on a peat block in a peat garden or raised bed.

INDOOR CULTIVATION

Most of the more vigorous species and cultivars can be grown relatively easily indoors on a window-sill if no glasshouse is available, indeed these charming plants are often called 'Window-sill Orchids'. The kitchen or bathroom window-sills are the best site for plants which prefer a moist atmosphere and plenty of fresh air. Their pseudobulbs should be kept in a garage or similar place that is just frost-free for the duration of their resting period (approx. November–January). A refrigerator at 3–5°C is ideal and late flowering can be readily achieved by leaving plants for a longer period at such low temperatures.

They can then be repotted and brought into the house and placed in an east- or west-facing window to initiate growth and flowering. When flowering has finished, the pans can be removed to the garden at the end of May, or after all danger of frost is past. The pans are best stood in a sheltered, shady spot in the garden to make their summer growth. However, they must be fed regularly and watered when dry to obtain the maximum-size bulbs. When they go dormant, or if there is a danger of frost in the autumn, they must be moved to a frost-free place again. To ensure a moist atmosphere the pans should be placed on moist gravel in a tray; however, care must be taken to ensure that the bases of the pans are not standing in water. Compost, feeding and rest period routines should be followed as previously described.

FLOWERING TIMES

All the commonly grown *Pleione* species and cultivars flower between February and April in cultivation. However, it is possible to have plants in flower from late September through until June if the selection is increased.

Pleione praecox is the first, coming into flower in late September or early October. This is quickly followed by *P. maculata* and then their hybrid, *P.* × *lagenaria*. The recently produced hybrids between autumn- and spring-flowering species, such as *P.* Tarawera, conveniently fill the gap in late December and January before the earliest of the spring species, *P. humilis*, begins to flower in late January or early February.

The earliest flowering clones of *P. formosana* such as 'Snow White' and hybrids like *P.* Eiger, which have *P. humilis* in their parentage, come into flower during late February and March.

Many species and cultivars flower during April and May. *Pleione hookeriana* completes the flowering season in late May and early June.

Pleiones are generally regarded as having solitary flowers, but most species and cultivars are capable of producing two, or occasionally even three, flowers per stem. Individual flowers do not last very long, fourteen days being a good average, though some hybrids will last for twenty-one days or more.

Although it is not generally recognized, several species produce fragrant flowers, this being most strongly developed in the autumn-flowering species. *Pleione praecox* has a striking primrose scent, whilst *P. maculata* and *P.* × *lagenaria* have scents reminiscent of sweet apples. Amongst the spring-flowering species several clones of *P. formosana* are faintly scented, as are *P. forrestii* and *P. speciosa*. Fortunately these species pass on their fragrance to their hybrids and some clones of the following grexes are appreciably fragrant—Rakata, Shantung, Soufrière, Stromboli, Tolima and Vesuvius.

PROPAGATION

Pleiones propagate readily by vegetative means, increasing three- or four-fold every year from mature pseudobulbs. All the clones of *P. formosana*, *P. speciosa* and *P. limprichtii* can produce two flowering-sized pseudobulbs a year from each of the previous season's flowering pseudobulbs. On the other hand *P. bulbocodioides*, *P. forrestii* and *P. hookeriana* will usually produce only one new flowering-sized pseudobulb. However, all these species are also capable of producing two or three small bulbils at the apex of the old shrivelled pseudobulb, although this may vary from season to season.

The bulbils are best grown on in trays or shallow pans. They should be scattered on the surface of the fine compost (see p. 22) together with a dusting of bone-meal and then completely covered by a thin layer of fine compost. Water should be given sparingly at first but increased as the new leaves appear. Bulbils will take two or three years to reach flowering size. Hybrids generally increase in much the same manner as their parents.

Pleiones can also be raised fairly readily from seed. The recent popularity of new hybrids is a reflection of this fact. Fruit-capsules mature after about 220 days. Seed can be sown in sterile conditions on a nutrient medium. This is best done by a laboratory specializing in this type of work, though it can be attempted even in a kitchen if certain precautions are followed; a fuller account of micropropagation techniques suitable for orchid seeds is given by Thompson (1977) and Arditti (1982).

Following successful germination, the seedlings are replanted on to another nutrient medium in a glass flask and are allowed to grow until they have produced tiny pseudobulbs. This usually coincides with autumn when they can be removed from the controlled environment of the laboratory to that of the glasshouse. They will then begin to go dormant and should be allowed a rest period, although the temperature should not be allowed to fall below about 3°C. In early January the tiny pseudobulbs should be removed from their flasks whilst still dormant. Clumps of five or six young pseudobulbs are pricked out into trays. The less they are disturbed at this stage the better is the survival rate.

The compost for the seedlings is the same as for the more mature pseudobulbs outlined previously. The covering of fine compost should include extra chopped sphagnum moss. Water should be given very sparingly until the new leaves have shown above the compost. Once active growth has commenced, plants can be fed regularly to build up the pseudobulbs to a good size in the first season. At the end of the year they can be separated and treated as already described for bulbils. Flowers will generally be produced four years after the original cross was made.

PESTS AND DISEASES

There are several pests and diseases which can afflict pleiones. Many of these stem from poor cultivation in the first place, so that correct watering, the use of well-drained composts, regular feeding and allowing a proper dormant season can help to reduce some of the problems.

Full accounts of orchid pests and diseases and their prevention are given in the excellent account of Williams (1980) and in the American Orchid Society's *Handbook on orchid pests and diseases* (revised edition 1986).

Bumble Bees. These can prove a nuisance by pollinating flowers which quickly wither as a result. They can be kept out of glasshouses by covering the ventilators with muslin.

Greenfly (Aphis). The most common pest and most noticeable early in the season, when they attack flower-buds. Flowers can become deformed as a result. Regular sprays with a systemic insecticide will keep them under control.

Mealy Bugs. These may prove a problem later in the season and can be controlled with a systemic insecticide.

Mice. Hungry mice may chew or sometimes decimate pseudobulbs during the winter and will need to be controlled. Unfortunately mice always seem to go for the best and rarest plants!

Red Spider Mites. During hot dry periods these may prove a problem, however, if the atmosphere is kept moist then much of the problem can be forestalled. Red spider mites generally attack the undersurfaces of the leaves and once their presence has been observed plants must be sprayed with a suitable insecticide.

Sciarid Fly. These tiny grey-black flies appear on the surface of the compost during the daytime, especially during hot weather. The 'grubs' feed on decaying humus in the compost but do little real damage. They can be controlled by regular spraying with a suitable insecticide.

Slugs and Snails. These common pests may destroy young shoots and flower-buds. A proprietary slug-killer will keep them in check.

Vine Weevils. The creamy white or pale brown 'grubs' of vine weevils burrow into the pseudobulbs, eventually leaving only a thin shell. They are not often detected and are most often seen on imported pseudobulbs and can be controlled by dusting the compost with gamma-HCH or, alternatively, using it as a soil drench in a liquid form.

Overwatering or poorly drained composts can lead to root-rot. Likewise, pseudobulbs stored badly and not given a sufficient rest period, or physical damage to plants, may also cause rotting. Once pseudobulbs have started to rot it is generally too late to save them. If the rot is confined to the roots then a fungicide can often reduce the problem. The systemic fungicide Benlate is perhaps the best to use though it must be used in strict accordance with the maker's instructions. Too strong a solution may cause premature dormancy.

Sometimes leaf-tips may turn brown and die during the growing season. Occasionally up to half the leaf surface may be affected. The cause of this is unknown, though correct culture and regular feeding throughout the growing season seem to reduce the problem. Plants that produce a strong root system early in the season are rarely affected by leaf-tip necrosis.

TAXONOMIC TREATMENT

Despite its small size *Pleione* presents the taxonomist with several problems which can be divided more or less into those of a practical and those of a biological nature. Among the former is the problem of reconstituting dried and pressed (and often badly pressed) thin-textured flowers of herbarium specimens without damaging them unduly. In particular, it is often difficult to spread out the lip where the callus hairs or lamellae are stuck to the opposite side of the lip or to another floral segment. A far more serious problem is that posed by the absence or destruction of type material. This is a particular problem with Schlechter and Kraenzlin species where the types were destroyed in Berlin during the Second World War, and no isotypes have yet been located. In such cases, the original description is often all that remains for the taxonomist to study.

The biological problems are of a more universal nature and involve such phenomena as hybridization in the wild, polyploidy and variation within a widespread species. Natural hybridization has been confirmed in both *P.* × *lagenaria* and *P.* × *confusa*, but may be more widespread than is currently appreciated. The opportunities for hybridization probably exist in any locality where two or more species are to be found flowering at the same time. Even spatial, altitudinal or ecological separation may not necessarily be a barrier because pollinating insects can often cover considerable distances. When several species are apparently sympatric (i.e. they have the same or overlapping geographical distribution, as in Yunnan), the chances of hybridization must be relatively high.

Polyploid clones have been demonstrated in *P. speciosa*, *P. limprichtii* and *P. bulbocodioides* and have undoubtedly led to confusion in this complex of closely allied species. The occasional presence of clones with very large flowers in *P. hookeriana* may also be attributable to polyploidy.

The most difficult problem is probably that of infraspecific variability. It has been known for many years that some species such as *P. praecox* and *P. humilis* are relatively variable both in flower size and coloration. Others such as *P. maculata* and *P. albiflora* appear to be more uniform. The introduction of selected clones into cultivation may lead growers to erroneous conclusions about the significance of these differences. Hunt & Vosa (1971) argued that, although some clones of the 'bulbocodioides complex' in cultivation were readily distinguishable from others, these did not warrant recognition at specific rank, but rather reflected the variability of *P. bulbocodioides*. As explained later we have not followed their treatment here but do recognize that their arguments may be valid in some cases.

Generic Description

Pleione D. Don, Prodr. Fl. Nepal. 36 (1825); Pfitzer, Morph. Stud. Orchid. Blüte 115, t.57 (1886) & in Engl. & Prantl, Pflanzenf. 2, 6: 126 (1889); Rolfe in Orchid Rev. 11: 289–92 (1903); Pfitzer & Kraenzl. in Engl., Pflanzenr. (IV.50) Coelogyninae 119 (1907); Schltr., Die Orchideen, ed. 1: 151–3 (1914), in Feddes Repert. Beih. 4: 61 & 185–6 (1919) & 12: 346 (1922); Hunt & Vosa in Kew Bull. 25: 423–32 (1971); Butterfield & Hunt in Plantsman 1: 112–23 (1979); Harberd in Brickell et al., Petaloid Monocotyledons 171–81 (1980); Wimber & Cribb in Plantsman 3: 178–88 (1981). Types: *P. humilis* (J.E. Sm.) D. Don & *P. praecox* (J.E. Sm.) D. Don.

Gomphostylis Wall. ex Lindl., Gen. Spec. Orch. 43 (1830).

Coelogyne Lindl. sect. *Pleione* (D. Don) Lindl., Gen. Spec. Orch. 43 (1830) & Fol. Orchid. 37 (1854); Reichb. fil. in Walpers, Ann. Bot. 6: 234 (1861); Pfitzer, Vergl. Morph. d. Orchid. 152, 154 (1882); Benth. & Hook. fil., Gen. Pl. 3: 519 (1883); Hook. fil., Fl. Brit. India 5: 828 (1890).

Coelogyne Lindl. sect. *Gomphostylis* (Wall. ex. Lindl.) Endl., Gen. 190, n. 13476 (1837).

DESCRIPTION. *Dwarf herbs, terrestrial, lithophytic or epiphytic. Pseudobulbs* annual, often clustered, ovoid, conical, pyriform or barrel-shaped, ± covered by sheaths when young, 1- or 2-leaved at the apex. *Leaves* deciduous, thin-textured, erect to arcuate-spreading, plicate, mostly shortly petiolate. *Inflorescence* basal, 1 to several per pseudobulb, erect, usually 1- but often 2-flowered, produced either before or after the leaves (with the leaves in *P. hookeriana* and *P. scopulorum*). *Flowers* often showy, sometimes fragrant, white or pink to rose-purple or magenta, more rarely yellow, marked with yellow, red or brown on the lip. *Sepals and petals* free, ± spreading. *Lip* (labellum) obscurely 3-lobed to entire, sometimes fused at the base to the column, apical margins erose to lacerate; callus consisting of 2 to several lines of lamellae or hairs along the veins of the upper surface of the lip.

Column slender, slightly arcuate, winged above, entire, erose or denticulate at the apex.

DISTRIBUTION. About 16 species widespread from C Nepal eastwards to Taiwan and from C China south to SE Burma, northern Thailand and Laos. MAP 1, below.

SECTIONS. Two sections are currently recognized. Sect. *Dictyopleione* Pfitz. includes the autumn-flowering species which possess turbinate, 2-leaved pseudobulbs which are mottled or warted, inflated or warted sheaths at the base of the inflorescence and a lip-callus consisting of papillate lines—no. 1–3 in the text. Sect. *Pleione* includes all the spring-flowering species which possess ovoid to flask-shaped pseudobulbs with a single leaf (2-leaved in *P. scopulorum*), smooth sheaths at the base of the inflorescence and a lip with either a hairy or lamellate callus—no. 4–16 in the text.

Pleione hookeriana and *P. scopulorum* currently in sect. *Pleione* differ from the other species in the section by having long scapes and small flowers with a lip broader than long. Furthermore, they also differ in seed morphology, the seeds more closely resembling those of some *Coelogyne* and *Bletilla* species. These two *Pleione* species may well warrant a separate section of their own within the genus.

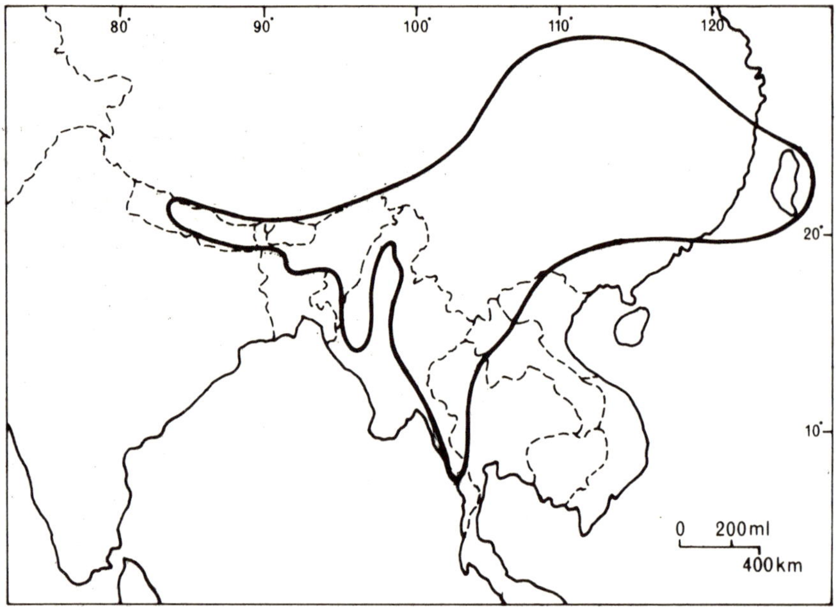

Map 1. Distribution of the genus *Pleione*.

Plate 5

Pleione coronaria CHRISTABEL KING

Plate 6

Pleione forrestii RODELLA PURVES

Key to Species and Natural Hybrids of Pleione

1. Pseudobulbs 2-leaved 2
   Pseudobulbs 1-leafed 5

2. Plants spring-flowering; pseudobulbs ovoid-conical; peduncle elongate, not covered by sheaths; callus consisting of 7 lacerate lamellae 5. **P. scopulorum**
   Plants autumn-flowering; pseudobulbs turbinate; peduncle short, covered entirely by sheaths; callus consisting of 3–7 papillate lines 3

3. Sheaths inflated, not warty; sepals and petals white 1. **P. maculata**
   Sheaths warty, but not markedly inflated; sepals and petals pink to rose-purple (rarely white) 4

4. Lip marked with purple blotches on apical margin; callus not reaching the apex of the lip; pseudobulbs green .... 2. **P. × lagenaria**
   Lip without blotches on the margin; callus reaching the apex of the lip; pseudobulbs mottled with reddish brown or purplish 3. **P. praecox**

5. Callus of lip consisting of 4–7 lines of hair-like outgrowths 6
   Callus of lip lamellate; lamellae margins entire, denticulate, erose or lacerate 8

6. Flowers small, lacking a saccate base to the lip; sepals and petals mostly less than 35 mm long; lip broader than long, somewhat reniform in outline when flattened; column short, 15–22 mm long; plants often stoloniferous .... 4. **P. hookeriana**
   Flowers medium-sized to large, with a definite saccate base to the lip; sepals and petals usually more than 35 mm long; lip longer than broad, somewhat obovate in outline; column 26 mm long or more; plants never stoloniferous 7

7. Lip pure white or with a central stripe of red or brown, callus with 5 lines of hairs; sepals mostly more than 44 mm long and the lip more than 40 mm long; column 30–40 mm long .... 6. **P. albiflora**
   Lip white, flecked with crimson or yellow-brown all over, callus with 6 or 7 lines of hairs; sepals 33–42 mm long and the lip 32–40 mm long; column 24–32 mm long .... 7. **P. humilis**

8. Callus-lamellae with entire margins 9
   Callus-lamellae with denticulate, erose, lacerate or undulate margins 10

9. Flowers pale yellow to orange-yellow (rarely white) with red or brown spotting on the lip; column 26–32 mm long; peduncle 20–50 mm long 9. **P. forrestii**
Flowers pale lavender to rose-pink (rarely white) with red or purple spotting on the lip; column 18–22 mm long; peduncle 7–15 cm long 11. **P. yunnanensis**

10. Flowers pale to primrose-yellow, spotted with red on the lip 10. **P. × confusa**
Flowers white, pink or rose-purple, spotted with brown, yellow or purple on the lip 11

11. Lip-callus of 6 (possibly 7) lamellae; mid-lobe of lip obflabellate, truncate, weakly denticulate; lip marked with red or purple dots at the apex and between the callus-lamellae; mature pseudobulbs with a distinct raised brown ring at the apex .... 8. **P. coronaria**
Lip-callus of 2–5 lamellae; mid-lobe of lip not obflabellate, dentate to lacerate, usually deeply emarginate; brown or purple spotting on lip at the apex and/or on the side-lobes; mature pseudobulbs usually lacking a marked apical corona of tissue 12

12. Flowers plain white, lacking any marking on the lip or markings very pale 13
Flowers pink, rose-purple or magenta, variously spotted with brown or purple on the lip 14

13. Lamellae of lip 5, unbroken, but with the margin lacerate into hair-like divisions; lip-margin coarsely lacerate; peduncle 8–12 cm long 12. **P. grandiflora**
Lamellae of lip 2–5, broken, but with ± erose margins; lip-margin fimbriate; peduncle usually less than 8 cm long ... 14. **P. formosana**

14. Lip almost circular in outline when flattened, 25–40 mm long; sepals and petals 28–36(–40) mm long; column 25–30 mm long 15. **P. limprichtii**
Lip obovate or flabellate in outline when flattened, more than 32 mm long; sepals and petals 33–64(–70) mm long; column usually more than 30 mm long 15

15. Lip markedly hook-shaped in side view, the apical third strongly deflexed, red spotting on lip confined to apical part of mid-lobe and often confluent; lamellae of lip 2 or 4, the one on the mid-vein absent, finely and evenly erose; column 35–45 mm long .... 16. **P. speciosa**
Lip not markedly hook-shaped in side view, spotting on lip on side- and mid-lobes; lamellae of lip usually 3–5, the one on the mid-vein usually present; column 27–36 mm long 16

16. Sepals and petals rose-pink to bright magenta, 33–50 mm long; lip similar in colour to sepals and petals, spotted with dark red-purple, 32–45 mm long, 25–35 mm wide; lamellae of lip unbroken, lacerate especially towards the apex .... 13. **P. bulbocodioides**

Sepals and petals pale to dark pink, 45–58 mm long; lip whitish to pink on margins, spotted with pale purple to yellowish brown, 50–55 mm long, 40–45 mm wide; lamellae of lip broken, erose or denticulate 14. **P. formosana**

1. PLEIONE MACULATA

There are only two autumn-flowering species of *Pleione*, *P. maculata* and *P. praecox*. *Pleione maculata* is a relatively uniform species and is readily distinguished from *P. praecox* by its small flowers which are white, with the lip marked with purple blotches on its margins, and by the broad bract with overlapping edges. Vegetatively it may also be distinguished by its green unmarked pseudobulbs and the smooth inflated sheaths that enclose the peduncle.

Pleione maculata was figured in *Curtis's Botanical Magazine* (t. 4691, 1852) from specimens sent to Kew by Simons and probably collected in the Khasia Hills of Assam. Unfortunately it has not proved an easy plant to grow and its persistence in cultivation owes much to its regular reintroduction from India. It is regrettable that wild plants are still extensively gathered and sold.

Plants of *P. maculata* from Thailand appear to have consistently smaller flowers than Indian material. The species is relatively restricted in distribution in Thailand but plants from there occasionally appear in cultivation. One, collected in north Thailand by David Menzies and David Du Puy in 1983, which flowered at Kew in early January 1984, had flowers half the size of Indian plants flowering at the same time.

In cultivation, flowers are usually produced during October and November, rather later than *P. praecox*. During the winter, plants need to be kept warmer than the other species, 3–5°C being ideal. In the early spring as root and leaf growth commences, watering must be undertaken with a good deal of caution as the roots can rot very easily at this stage. Once active growth is resumed then the watering can be increased.

At the Royal Botanic Garden in Edinburgh, specimens of *P. maculata* from Bhutan have been successfully grown tied to a piece of cork covered with moss, hanging vertically in a shaded position where the minimum temperature does not drop below 18°C during the growing season.

In the wild, *P. maculata* is far less widespread than its close cousin *P. praecox*, being generally a plant of lower altitudes. However, in the Khasia Hills of Assam the two may be found growing in close proximity between 1200 and 1600 m. The existence of a putative hybrid, *P.* × *lagenaria*, from this region testifies at least to some overlap in flowering time of these two species in their native habitats.

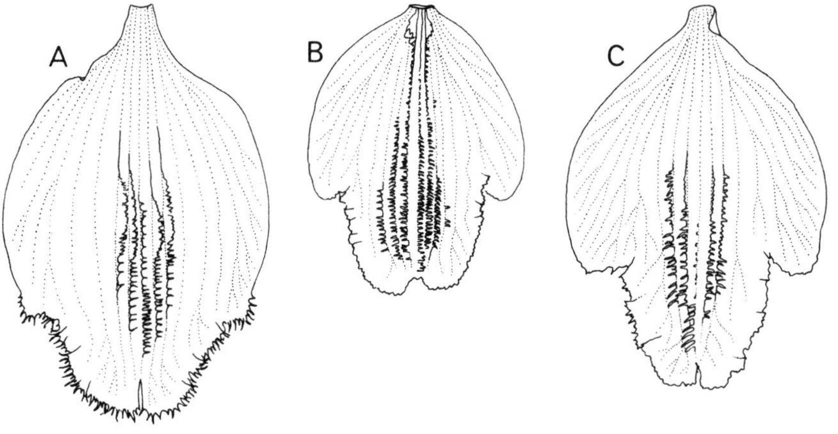

Figure 1. Lips of *Pleione*, × 1. A, *P. praecox*, Kew Spirit Colln. 45416; B, *P. maculata*, Kew Spirit Colln. 20299; C, *P.* × *lagenaria*, cult. Kew ex Butterfield.

Pleione maculata (Lindl.) Lindl. in Paxt., Fl. Gard. 2: 5, t.39 (1851); Hunt & Vosa in Kew Bull. 25: 430 (1971); Cribb et al. in Curtis's Bot. Mag. 184: 106, t.860 (1983).

Coelogyne maculata Lindl., Gen. Sp. Orch. Pl. 43 (May 1830) & in Wall., Pl. As. Rar. 1: 45, t.53 (July 1830). Type: India, Pundua, *Wallich* 1964 (holotype K!).

Pleione diphylla Lindl. in Paxt., Fl. Gard. 2: 66 (1851). Type: India, Khasia, *Griffith* 29/138 (holotype K!).

Coelogyne diphylla (Lindl.) Lindl., Fol. Orch. Coelog. 15 (1854).

C. arthuriana Reichb. fil. in Gard. Chron., ser. 2, 15: 40 (1881). Type: origin unknown, cult. *Veitch* (holotype W!).

Pleione maculata (Lindl.) Lindl. var. *virginea* Reichb. fil. in Gard. Chron., ser. 3, 2: 682 (1887). Type: origin unknown, cult. *W.H. Scott* (holotype W!).

P. maculata (Lindl.) Lindl. var. *arthuriana* (Reichb. fil.) Rolfe ex Kraenzl. in Engl., Pflanzenr. (IV.50) Coelog. 128 (1907).

Gymnostylis candida Wall. ex Pfitz. in Engl., op. cit. 127 (1907).

Description. *An epiphytic herb. Pseudobulbs* turbinate, beaked, 10–30 mm long, 10–15 mm in diameter, covered by remains of old sheaths, 2-leaved at the apex. *Leaves* elliptic-lanceolate to oblanceolate, acute, 10–25 cm long, 1.5–3.5 cm wide. *Inflorescence* 1-flowered, appearing before the leaves. *Peduncle* entirely covered by inflated, smooth sheaths. *Flowers* suberect to spreading, 40–60 mm across, fragrant; sepals and petals creamy white, occasionally lightly streaked with pink, the lip white with a yellow centre, purple blotches on the apical margin and white lamellae. *Bract* nearly circular when flattened, 17–30 mm long, obtuse, cucullate. *Dorsal sepal* oblong-lanceolate, 30–40 mm long, *c.* 8 mm wide, obtuse. *Lateral sepals* broadly lanceolate-falcate, 30–40 mm long, *c.* 9 mm wide, acute. *Petals* oblanceolate, slightly falcate, 30–42 mm long, *c.* 8 mm wide, acute. *Lip* 3-lobed, oblong, 25–35 mm long, 20–25 mm wide; side-lobes obscure; mid-lobe about one-third the length of the lip, emarginate with an undulate erose margin; callus consisting of 5–7 papillate lines, extending to the apex of the lip. *Column* 17–20 mm long, erose-dentate at the apex. 2n = 40. PLATES 1, 15D; FIGURE 1B, p. 37.

DISTRIBUTION. Bhutan, Burma, SW China (Yunnan), India (Assam, Manipur & Sikkim), and N Thailand; altitude 600–1600(–2600) m. MAP 2, below.

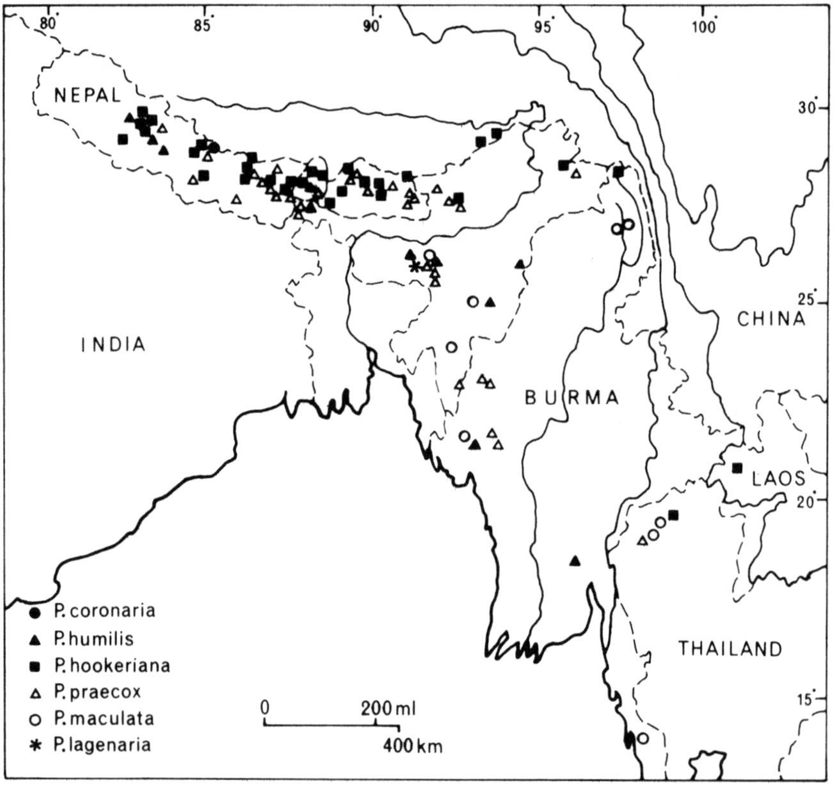

Map 2. Distribution of *Pleione* species.

2. PLEIONE × LAGENARIA

Pleione lagenaria was figured in *Curtis's Botanical Magazine* (t.5370, 1863) but has remained relatively rare in cultivation ever since. However, judging from the herbarium material examined at Kew and Edinburgh it has possibly been in cultivation more or less continuously since its original introduction in 1851. John Day (1866) in his scrapbooks states that *P. lagenaria* is a 'freer bloomer' and is 'soon propagated', thus lending weight to the view that this lovely plant may have survived in cultivation for a considerable period.

Reference has been made to the possible hybrid origin of *P. lagenaria* under *P. maculata* (p. 36).

Pleione lagenaria was first described by John Lindley in 1851, based on a specimen collected in the Khasia Hills, Assam, by Thomas Lobb for Messrs Veitch & Sons. Lindley treated this plant as a distinct species, though stating that it was received mixed with *P. maculata*. Both *P. maculata* and *P. praecox* are common in the Khasia Hills where their altitudinal ranges overlap, and they are often found growing together there, as indeed they may be elsewhere over their range. Their flowering periods also overlap and it is quite possible that they sometimes hybridize in the wild. The situation is analogous to that of another hybrid, *P. forrestii* × *P. albiflora*, introduced from the wild and grown for many years as *P. forrestii* before its correct identity was established (Wimber & Cribb, 1981).

Hunt & Vosa (1971) relegated *P. lagenaria* to synonymy within *P. praecox* without explanation other than that provided by their key, which circumscribed *P. praecox* as having warty basal sheaths on the flowering stem, and flowers with a lip 30–45 mm long and 35–55 mm wide, bearing lamellae only two-thirds the length of the lip. They contrasted it with *P. maculata* in which the basal sheaths lack warts and the flowers are smaller, the lip 20–30 mm long and 25–35 mm wide, bearing lamellae extending to the apex.

Although *P. lagenaria* possesses the warty basal sheaths of *P. praecox* and a similar lip-callus, it also resembles *P. maculata* in many features. Thus, its basal sheaths are slightly inflated, its lip is similar in colour to *P. maculata* and its flowers are smaller than those of typical *P. praecox*. An analysis of *P. lagenaria* reveals that in many features it is intermediate between *P. praecox* and *P. maculata* and this is clearly shown in the Table overleaf. Therefore, we suggest that this plant is a natural hybrid between *P. maculata* and *P. praecox*.

Table: Comparison of *P. praecox*, *P.* × *lagenaria* and *P. maculata*

Character	*P. praecox*	*P.* × *lagenaria*	*P. maculata*
Pseudobulbs	green and purple mottled	green and purple mottled	green
Inflorescence height	8–13 cm	8–11 cm	5–8 cm
Inflorescence sheaths	pustulate, non-inflated	pustulate, slightly inflated	smooth, inflated
Sepal and petal colour	white to rose-purple	pink to rose-purple	white
Dorsal sepal length	50–70 mm	44–50 mm	30–40 mm
Lip-margin	unblotched	heavily blotched	heavily blotched
Lip shape	elliptic	oblong-elliptic	oblong
Lip length	40–50 mm	32–40 mm	25–35 mm
Lip-base	saccate	not saccate	not saccate
Callus	3–5 papillate lines not reaching apex	5 papillate lines not reaching apex	5–7 papillate lines reaching apex
Column length	35–45 mm	28–30 mm	17–20 mm
Ovary length	16–32 mm	15–21 mm	15–24 mm
Bract shape	obovate, acute	obovate, acute	elliptic, obtuse
Bract length	*c.* 30 mm	28–32 mm	17–30 mm

Pleione × lagenaria Lindl. in Paxt., Fl. Gard. 2: 5, t.39 (1851); Cribb et al. in Curtis's Bot. Mag. 184: 107, t.860 (1983). Type: Paxt., Fl. Gard. 2: 5, t.39.
Coelogyne lagenaria (Lindl.) Lindl., Fol. Orch. Coelog. 15 (1854).

Description. *An epiphytic or lithophytic herb. Pseudobulbs* turbinate, *c.* 26 mm long, *c.* 24 mm in diameter, covered by fibrous remains of old sheaths, 2-leaved at the apex. *Leaves* oblanceolate, acute, up to 32 cm long and 4.8 cm wide, petiolate. *Inflorescence* 1 or 2 per pseudobulb, 6–8 cm tall. *Peduncle* 2–3 cm long, covered by *c.* 3, somewhat inflated, pustulate sheaths. *Bract* obovate, acute, 28–32 mm long, cucullate, covering the ovary. *Flower* suberect-spreading; sepals and petals pink to rose-purple; lip white with a yellow central area and purple blotches around the margin. *Sepals* linear-lanceolate, 44–50 mm long, 7–9 mm wide, acute. *Petals* linear, 42–50 mm long, 3–5 mm wide, acute. *Lip* 3-lobed, 32–40 mm long, 30 mm wide; side-lobes erect; mid-lobe longer than the side-lobes, subquadrate, emarginate, dentate; callus consisting of 5 longitudinal rows of short papillae along the central veins and running on to the mid-lobe of the lip. *Column* 28–30 mm long. 2n = 40. PLATE 1; FIGURE 1C, p. 37.

DISTRIBUTION. Assam and possibly SW China (Yunnan); altitude 1500–2500 m. MAP 2, p. 38.

3. PLEIONE PRAECOX

Pleione praecox is closely related to *P. maculata*, the only other autumn-flowering *Pleione* species. It can be readily distinguished, however, by its larger, distinctively coloured flowers which have a saccate base to the lip, by its narrower bract and purple-spotted, pustulate sheaths on the peduncle and, furthermore, by the characteristic purple-mottled pseudobulbs.

Pleione praecox is a far more variable species than *P. maculata* in both flower size and colour. The sepals and petals vary from white flecked with lilac, to rose-purple or lilac-purple. The lip is similarly variable from white to purple, usually with a central yellow streak, but more rarely with purple blotches on the side-lobes.

Two particularly attractive varieties were introduced into cultivation from Nepal by C. Bailes in 1984, one with dark rose-pink or purple flowers and the other a striking albino which has been given the cultivar name 'Everest'. The latter regularly produces two flowers on each inflorescence.

In the wild *P. praecox* is known from higher altitudes than *P. maculata*, so in cultivation the plants need to be kept cold during the winter. The flowers appear just as growth is finishing, so it is possible to have the pink-flushed autumn leaves with the flowers as they open. Well-grown pseudobulbs may produce two flowering stems, and sometimes (at least in cultivated specimens) each stem may bear two flowers.

Detailed accounts of the habitat of *P. praecox* in Nepal are given by Butterfield & Bailes (1986) and Kretz (1987). They found it in abundance on rocky slopes by the roadside at between 1375 and 2000 m. The plants were found in colonies on moss-covered and often vertical rock-faces and occasionally in humus pockets and at the base of shrubs or amongst rocks. The plants were in full flower in October and November.

Pleione reichenbachiana (T. Moore & Veitch) B.S. Williams was described in 1868, based on plants collected for Messrs Veitch by Col. Benson in the Arracan Mountains of Burma. This plant differs from typical *P. praecox* from Assam, Nepal and Sikkim in having slightly smaller flowers with a white lip spotted with purple on the sides, in lacking the central yellow streak to the lip and in possessing three, rather than four or five, lamellae. Similar plants have been more recently collected in north Thailand by H.B.G. Garrett.

Examination of the available herbarium material suggests that forms intermediate beween *P. reichenbachiana* and *P. praecox* exist. As a result, in this account, we are following Hunt & Vosa (1971) in treating the former as a synonym of *P. praecox*. Likewise, *P. birmanica* (Reichb. fil.) B.S. Williams collected in Burma by W. Boxall for Messrs Low agrees closely with *P. reichenbachiana* judging from Reichenbach's original description. It has been reduced to synonymy within *P. praecox*.

Pleione praecox from north India was first illustrated in *Curtis's Botanical Magazine* in 1850 (t.4496) under the name *Coelogyne wallichii* (another synonym and sphalm. for *C. wallichiana*).

Pleione praecox (J.E. Sm.) D. Don, Prodr. Fl. Nepal. 37 (1825); Cribb et al. in Curtis's Bot. Mag. 184: 110, t.861 (1983).

Epidendrum praecox J.E. Sm., Exot. Bot. 2: 73, t.97 (1806). Type: Nepal, *Buchanan-Hamilton* (holotype Hb. Linn.!).

Coelogyne praecox (J.E. Sm.) Lindl., Collect. Bot., sub. t.37 (1821).

C. wallichiana Lindl., Gen. Sp. Orch. Pl. 43 (May 1830) & in Wall., Pl. As. Rar. 1: 46, t.54 (July 1830). Type: India, Pundua, *Wallich* 1965 (holotype K!).

C. wallichii Hook. in Curtis's Bot. Mag. 76: t. 4496 (1850), sphalm. *C. wallichiana*.

Pleione wallichiana (Lindl.) Lindl., in Paxt., Fl. Gard. 2: 66 (1851).

Coelogyne praecox (J.E. Sm.) Lindl. var. *sanguinea* Lindl., Fol. Orchid. Coelog. 16 (1854). Type: Sikkim, *Hooker* 73 (holotype K!).

C. praecox (J.E. Sm.) Lindl. var. *wallichiana* (Lindl.) Lindl., loc. cit. (1854).

C. reichenbachiana T. Moore & Veitch in Gard. Chron., 1868: 1210 (1868). Type: Burma, Moulmein, *Benson* s.n. (not seen).

Pleione reichenbachiana (T. Moore & Veitch) B.S. Williams, Orch. Grow. Man. (ed. 4) 252 (1871).

Coelogyne birmanica (Reichb. fil. in Gard. Chron., ser. 2, 18: 840 (1882). Type: Burma, *Boxall* s.n. (holotype W!).
C. praecox (J.E. Sm.) Lindl. var. *tenera* Reichb. fil., op. cit. 20: 294 (1883). Type: origin unknown, cult. *Bull* (holotype W!).
Pleione concolor Hort. ex B.S. Williams, Orch. Grow. Man. (ed. 7) 681 (1894). Type: not known.
P. birmanica (Reichb. fil.) B.S. Williams, loc. cit. (1894).
P. praecox (J.E. Sm.) D. Don var. *birmanica* (Reichb. fil.) Grant, Orch. Burm. 167 (1895).
P. praecox (J.E. Sm.) D. Don var. *sanguinea* (Lindl.) Pfitz. in Engl., Pflanzenr. (IV.50) Coelog. 126 (1907).
P. praecox (J.E. Sm.) D. Don var. *candida* Pfitz., loc. cit. (1907). Type: not known.
P. praecox (J.E. Sm.) D. Don var. *alba* E.W. Cooper in Roy. Hort. Soc., Dict. Gard. 1606 (1951). Type: not known.
P. praecox (J.E. Sm.) D. Don var. *wallichiana* (Lindl.) E.W. Cooper, loc. cit. (1951).

DESCRIPTION. *An epiphytic or lithophytic herb. Pseudobulbs* turbinate, contracted abruptly above into a beak, 15–30 mm long, 10–15 mm in diameter, green mottled with reddish brown, or purplish, 2-(rarely 1-)leaved. *Leaves* elliptic-lanceolate to elliptic, acuminate, 15–26 cm long, 3–7 cm wide; petiole up to 6 cm long. *Inflorescence* 1(–2)-flowered, 8–13 cm tall, usually appearing after the leaves have fallen. *Peduncle* covered by *c.* 3 warty sheaths. *Bract* elliptic, acute, *c.* 30 mm long and 18 mm wide, longer than the ovary. *Flower* large and showy, white to rose-purple with yellow lamellae on the lip. *Sepals* narrowly oblong-lanceolate, acute, 50–70 mm long, *c.* 10 mm wide. *Petals* somewhat falcate, linear-lanceolate, acute, 50–70 mm long, *c.* 6 mm wide. *Lip* elliptic, 3-lobed, 40–50 mm long, *c.* 30 mm wide; mid-lobe emarginate, dentate-lacerate; callus consisting of 3–5 papillate lines running on to the middle of the mid-lobe. *Column* slender, 35–45 mm long, incurved, dentate at the apex. 2n = 40. PLATES 2, 15A, 19B; FIGURE 1A, p. 37.

DISTRIBUTION. Bhutan, Burma, SW China (Yunnan), N India (including Sikkim), Nepal and N Thailand; altitude 1200–3400 m. MAP 2, p. 38.

4. PLEIONE HOOKERIANA

Pleione hookeriana is one of the most widespread species in the wild, being found from central Nepal east to south China. It varies somewhat in flower colour, the sepals and petals ranging from white to rose-pink and the lip may sometimes be blotched with yellow, orange or reddish brown. Occasionally, specimens occur in the wild with flowers twice the normal size. Some collected by Adam Stainton at Solu Khola in east Nepal, under the number 4542, are of this type. These have the appearance of being tetraploid forms, although confirmation of this is lacking.

In its native habitats, *P. hookeriana* grows in thick moss as an epiphyte in mixed *Rhododendron/Tsuga* forest, but it also occurs on mossy rocks within the forest. It is often found growing in close proximity to *P. humilis*, though generally growing in large clumps up to a metre across rather than forming collars around the trunks and branches (Grey-Wilson, pers. comm.).

In general *P. hookeriana* grows at a higher altitude than most other *Pleione* species, often being covered by snow for at least part of the winter. In cultivation it tends to have a rather short growing season of about six months, so it is wise to keep the dormant pseudobulbs cold for as long as possible in the spring, only starting them into growth during late April or early May. Regular summer feeds are necessary to build up the pseudobulbs to a good size and the pans of plants benefit from a topping of live sphagnum moss. Because of the long dormant season it may prove necessary to dampen the compost occasionally during the winter to prevent the pseudobulbs from shrivelling. During the summer this species prefers plenty of air movement around the plants, so the pans need to be placed close to a fan or open ventilator. On the whole *P. hookeriana* is rather more difficult to grow than some of its cousins, neither increasing freely,

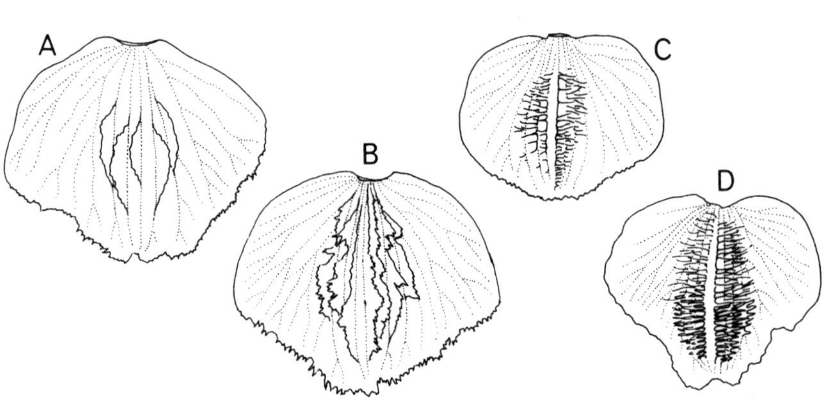

Figure 2. Lips of Pleione, ×1. A, *P. scopulorum*, Kingdon Ward 6283; B, *P. scopulorum*, Forrest 14230; C, *P. hookeriana*, Kew Spirit Colln. 34520; D, *P. hookeriana*, Kew Spirit Colln. 23217.

nor consistently producing flowering-sized pseudobulbs. However, it is a pretty species and well worth persevering with.

Pleione hookeriana is easily recognized by its small flowers in which the lip is usually wider than long and has seven lines of barbate hairs. Its closest ally is the rare and little-known *P. scopulorum* which is similar in flower size and shape, but differs in having lacerate lamellae on the lip and two-leaved, not one-leafed, pseudobulbs. *Pleione hookeriana* is the only species that produces stolons (Kretz, 1987). These are slender and sometimes several centimetres long.

A few albino clones of *P. hookeriana* have recently appeared in cultivation. One grown and flowered by one of the authors (IB) in 1987 had pure white flowers with golden yellow markings on the lip.

Pleione hookeriana (Lindl.) B.S. Williams, Orch. Grow. Man. (ed. 6) 548 (1885); Hunt in Kew Bull. 25: 428 (1971); Cribb et al. in Curtis's Bot. Mag. 184: 112, t.862 (1983).

Coelogyne hookeriana Lindl., Fol. Orch. Coelog. 14 (1854). Type: Sikkim, Darjeeling, *J.D. Hooker* 74 (holotype K!).

C. hookeriana Lindl. var. *brachyglossa* Reichb. fil. in Gard. Chron., ser. 3, 1: 833 (1887). Type: Sikkim, cult. *Lawrence* (holotype W!).

Pleione hookeriana (Lindl.) B.S. Williams var. *brachyglossa* (Reichb. fil.) Rolfe in Orch. Rev. 11: 291 (1903).

P. laotica Kerr in Journ. Siam. Soc. Nat. Hist. Suppl. 9: 235 (1933). Type: Laos, Pu Bia, *Kerr* 0977 (holotype K!).

DESCRIPTION. *An epiphytic or lithophytic herb. Pseudobulbs* conical to ovoid, 10–30 mm long, 5–15 mm in diameter, often clustered in large colonies, green or purple, 1-leafed, often producing stolons, up to 3 cm long or more, 1–1.5 mm in diameter. *Leaf* elliptic-lanceolate to oblanceolate, acute, 5–21 cm long, 1–4.6 cm wide; petiole 3–4 cm long. *Inflorescence* 7–14 cm long, 1-flowered, appearing with the young leaf. *Peduncle* slender, erect, 7–10.5 cm long. *Bract* ovate-elliptic, rounded at the apex, 10–15 mm long, 9–12 mm wide. *Flower* rather small; sepals and petals lilac-pink to rose or rarely white; lip white with a yellow disc and lamellae, spotted with purple or yellowish brown. *Dorsal sepal* oblong-lanceolate to oblanceolate, 20–32 mm long, 6–9 mm wide, acute. *Lateral sepals* falcate-lanceolate, 15–26 mm long, *c.* 10 mm wide, acute. *Petals* spreading, oblanceolate, 25–35 mm long, 5–7 mm wide, acute. *Lip* cordate to somewhat reniform (when flattened), obscurely 3-lobed, emarginate, 20–40 mm long, 25–40 mm wide, margin denticulate; callus consisting of 7 barbate lamellae. *Column* 15–22 mm long, broadly winged above, dentate at the apex. *Ovary* 7–17 mm long. *Fruit* ellipsoid, 15–19 mm long on a 15–16 mm long stalk. 2n = 40. PLATES 3, 15C; FIGURES 2C, 2D, p. 45.

DISTRIBUTION. Bhutan, Burma, China (S Xizang (Tibet) & Guizhou), NE India (Assam & Sikkim), Laos, C & E Nepal and N Thailand; altitude 2200–4200 m. MAP 2, p. 38.

5. PLEIONE SCOPULORUM

Pleione scopulorum occupies a somewhat isolated position in the genus. Like *P. praecox* and *P. maculata*, it is two-leaved but this is no indication of any close affinity, for unlike them, it is spring-flowering and has simple tubular sheaths, an ovoid rather than turbinate pseudobulb and small flowers with a distinctive lip and callus. In fact its affinities lie rather with *P. hookeriana* with which it shares small flowers in which the lip is wider than long, being widest towards the base. *Pleione scopulorum* differs, however, from *P. hookeriana* by having the leaves nearly fully developed at flowering time, by its two-leaved pseudobulbs, different flower colour and the lacerate lamellae of the lip-callus. In many ways *P. scopulorum* resembles a *Bletilla* and indeed was transferred, albeit mistakenly, to that genus by Schlechter.

In the wild it is one of the most restricted in distribution, known only from the adjacent parts of west Yunnan, south-east Xizang (Tibet), north Burma and with a solitary record from Naga Hills of Assam. It has been collected several times and by some of the most famous collectors such as Forrest, Farrer, Kingdon Ward and Rock but none of them introduced it and only in the last two or three years has it appeared in cultivation. This is an exciting development for it is an attractive plant with some highly desirable characteristics. The plants have a tall erect inflorescence with an attractive flower said to be bright scarlet-rose (Forrest) or vivid magenta (Farrer), with dark crimson spots on the lip which has a yellow or white centre and white lamellae. White-flowered specimens have been collected by both Farrer and Rock, but perhaps the most surprising and spectacular specimens were collected by Kingdon Ward in north Burma under the number 6823. These had bright sulphur-yellow flowers with orange spots and stripes on the lip but were otherwise quite typical of *P. scopulorum*. Such colour dimorphism is also known in the common European orchid *Dactylorhiza sambucina*.

Pleione scopulorum was introduced into cultivation from China late in 1985 or early in 1986, the first plants flowering in Europe in the spring of 1986. The few plants that have been seen to date have flowers not unlike those of *P. hookeriana* in size and attitude. Of these, two had rich rose-purple flowers with a lip with a yellowish base and white central area, sparsely marked with purple spots, while one also had yellow marks at the apex of the callus. The third clone was paler in colour with a pale pink lip with a few purple spots around the callus apex and with a yellow callus. All of these recent introductions have come from near the Mekong river in western Yunnan.

In the wild *P. scopulorum* grows in a variety of montane habitats; shaded situations on the grassy ledges of cliffs, in ravines on humus-covered boulders overhanging torrents, in pine forests and in open turf of the sub-alpine zone.

Pleione scopulorum W.W. Sm. in Notes Roy. Bot. Gard. Edinb. 13: 218 (1921); Hunt & Vosa in Kew Bull. 25: 429 (1971); Cribb et al. in Curtis's Bot. Mag. 184: 114 (1983). Type: China; Yunnan, *Forrest* 14230 (holotype E!; isotype K!).

Bletilla scopulorum (W.W. Sm.) Schltr. in Feddes Rep. Sp. Nov. 19: 375 (1924).

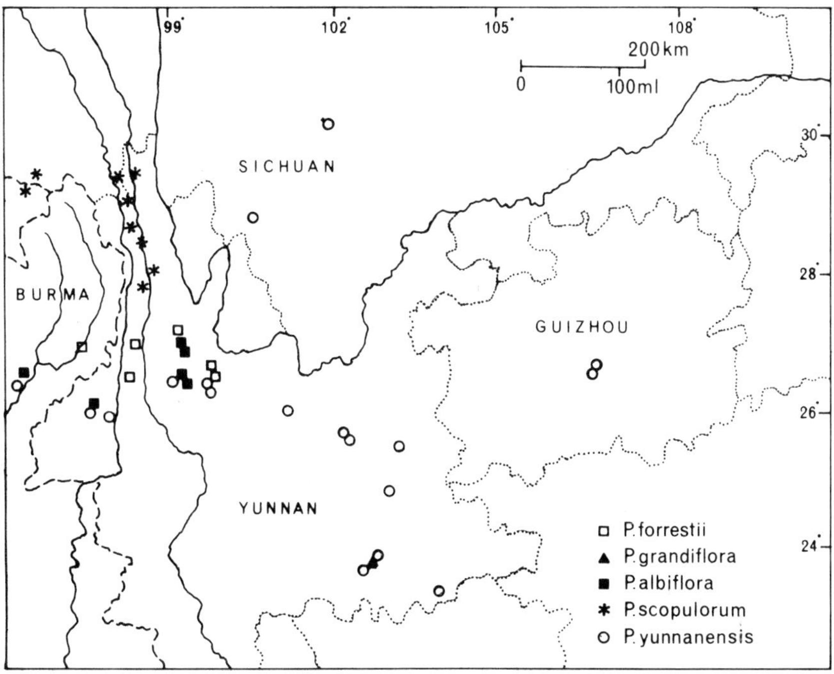

Map 3. Distribution of *Pleione* species.

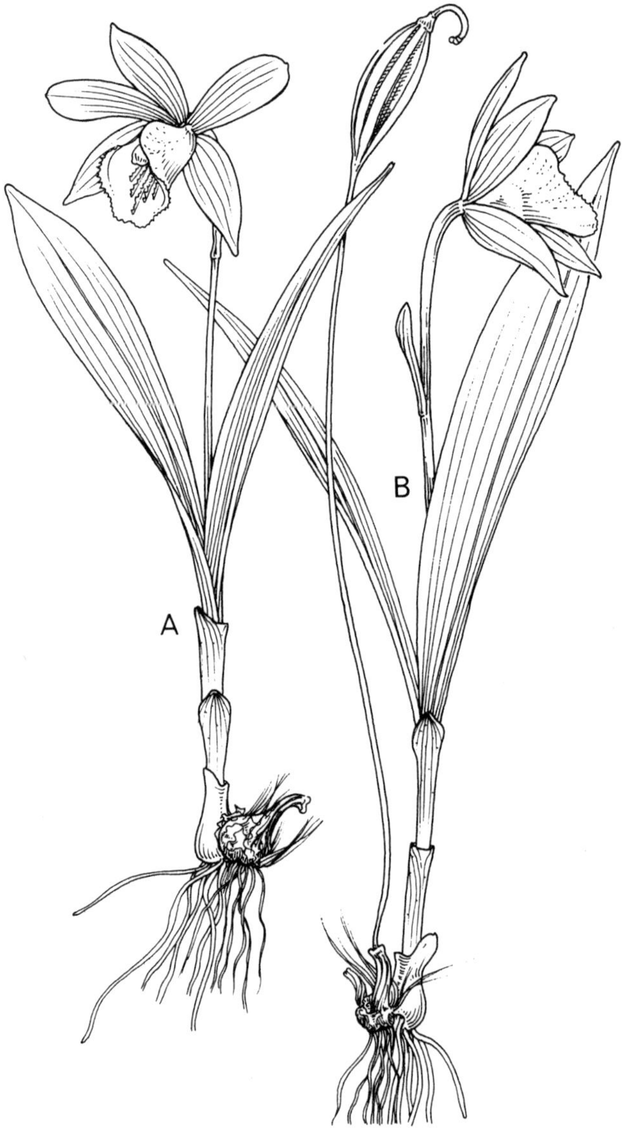

Figure 3. *Pleione scopulorum*, ×1. A, Forrest 14230; B, Forrest 19618.

Plate 7

Pleione × confusa

MARGARET STONES

Plate 8

Pleione yunnanensis CHRISTABEL KING

Plate 9

Pleione bulbocodioides CHRISTABEL KING after RODELLA PURVES

Plate 10

Pleione formosana 'Polar Sun' (top)
'Blush of Dawn' (middle)
'Oriental Grace' (bottom)

CHRISTABEL KING

Description. *A terrestrial or lithophytic herb* up to 30 cm tall. *Pseudobulbs* often oblique, ovoid to narrowly ovoid with a long slender neck, up to 25 mm long and 10 mm in diameter, covered with 3 or 4 loose sheaths when young, 2-leaved. *Leaves* developing at flowering time, lanceolate to oblanceolate, up to 12 cm long and 2 cm wide, acute. *Inflorescence* 1-(rarely 2- or 3-)flowered. *Peduncle* terete, up to 15 cm long. *Bracts* sheathing, oblanceolate, 19–36 mm long, obtuse to acute. *Flowers* rose-purple or rarely white or yellow, the lip spotted and streaked with dark purple. *Dorsal sepal* elliptic-lanceolate, 28–35 mm long, 7–9 mm wide, subacute. *Lateral sepals* obliquely elliptic, 21–32 mm long, 7–10 mm wide, apiculate. *Petals* oblanceolate, 28–31 mm long, 7–10 mm wide, subacute to obtuse, *Lip* obscurely 3-lobed, transversely elliptic-subreniform, wider than long, 18–30 mm long, 25–36 mm wide, the front margin denticulate; side-lobes erect; callus consisting of 7 high lacerate lamellae. *Column* arcuate, narrowly winged, 15–22 mm long, denticulate at the apex. *Ovary* 17–36 mm long. *Fruit* fusiform, 36–53 mm long. 2n = 40. plates 15E, 15F; figures 2A, 2B, 3, p. 45 and opposite.

Distribution. N Burma, SW China (SE Xizang (Tibet) & W Yunnan) and NE India (Assam); altitude 2800–4200 m. map 3, p. 47.

6. PLEIONE ALBIFLORA

Several collections at Kew, Edinburgh and the British Museum which have been identified by various authorities as *P. grandiflora* Rolfe do not agree at all closely with the type which Augustine Henry collected near Mengtze in south Yunnan in 1898. All these specimens come from north Burma and west Yunnan over 400 km from Mengtze. They differ from *P. grandiflora* in having a more slender pseudobulb, a shorter peduncle and a flower with a distinctly shaped lip which is saccate at the base, with four or five lines of long papillae on the apical half of the upper surface and a finely lacerate apical margin.

In many ways, these plants resemble large-flowered specimens of *P. humilis* and are undoubtedly more closely allied to that species than to *P. grandiflora*. However, they differ from *P. humilis* in several ways, sufficient for them to be recognized as a distinct species, *P. albiflora*; the flower is larger with a more prominent pouch at the base of the lip, the callus consists of four or rarely five lines of papillae along the apical half of the lip, the bract is broader and truncate and the pseudobulbs are generally conical and lack a noticeably narrow neck.

Pleione albiflora is rather variable in flower size and colour.

Specimens have been recorded with pure white flowers, more rarely with the sepals and petals lined with pale mauve, and the lip tinged with purple at the apex or with a bold crimson, brown or brownish yellow central stripe. In the wild it grows either as an epiphyte in moss on tree-trunks or on moss-covered rocks and cliffs in shady places.

The distributions of *P. humilis* and *P. albiflora* do not apparently overlap, the former being distributed from Nepal eastward to west-central Burma while the latter has a much more restricted distribution in the triangle of north Burma and adjacent west Yunnan. *Pleione albiflora*, rather than *P. grandiflora*, is undoubtedly one parent of the hybrid (*P.* × *confusa*, p. 58) grown for years under the name *P. forrestii*. It grows over the same distributional and altitudinal

Figure 4. *Pleione albiflora*, ×1, Forrest 4860.

range as the other parent, true *P. forrestii*, while *P. grandiflora* is known only from Mengtze, well ouside the range of *P. forrestii*.

Pleione albiflora Cribb & C.Z. Tang in Curtis's Bot. Mag. 184: 117 (1983). Type: China; Yunnan, Tali [Dali] Range, *Forrest* 15554 (holotype K!; isotype E!).

DESCRIPTION. *An epiphytic or lithophytic herb. Pseudobulbs* ovoid-conical with an elongated neck, 30–45 mm long, 8–18 mm in diameter, surmounted by a conspicuous scarious leaf-base 3–4 mm long, 1-leafed. *Leaf* falcate-lanceolate, acute, only known from immature specimens. *Inflorescence* 1-flowered. *Peduncle* short, 3–7(–9) cm long, covered, except near the apex, by 3 or 4 sheaths. *Bract* erect, obovate, 20–35 mm long, truncate or ± rounded at the apex. *Flowers* nodding, large, white, sometimes with crimson or brown markings on the lip, fragrant. *Sepals* subsimilar, narrowly elliptic, 44–54 mm long, 8–10 mm wide, obtuse. *Petals* oblanceolate, 45–54 mm long, 8–9 mm wide, obtuse or ± rounded at the apices. *Lip* obscurely 3-lobed, 40–57 mm long, 30–36 mm wide, saccate at the base to form a short spur 1–2 mm long, apical half of margin lacerate; side-lobes erect-incurved; callus consisting of 5 lines of long papillae in the apical half of the lip. *Column* slightly arcuate, 33–40 mm long, entire at the apex. FIGURES 4, 5C, opposite and below.

DISTRIBUTION. N Burma and SW China (W Yunnan); altitude 2400–3250 m. MAP 3, p. 47.

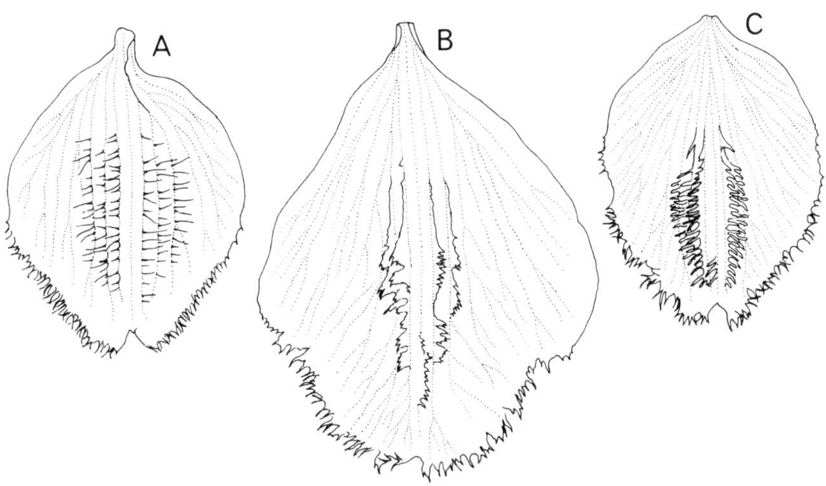

Figure 5. Lips of *Pleione*, ×1. A, *P. humilis*, Kew Spirit Colln. 17534; B, *P. grandiflora*, Henry 11116; C, *P. albiflora*, Forrest 24409.

7. PLEIONE HUMILIS

Pleione humilis is the earliest of the spring-flowering species, often being in full flower in cultivation by late January. It is a distinctive plant with one, or less frequently two, somewhat nodding, white flowers spotted with yellow-brown or crimson on the lip. Specimens with pink-flushed or pink-flecked sepals and petals have sometimes been recorded. Typical plants of this species were illustrated in *Curtis's Botanical Magazine* (t.5674) in 1867 based on specimens sent from Sikkim by Dr Anderson of the Calcutta Botanic Garden.

Specimens collected by Kingdon Ward on Mt. Victoria in the Chin Hills of west-central Burma which are still in cultivation have slightly larger flowers and more elongated pseudobulbs than the typical form; a pan of this was exhibited as *P. humilis K.W.* 21966 (in cultivation it is commonly known as 'Frank Kingdon Ward') at the Royal Horticultural Society in 1962 where it received a Certificate of Preliminary Commendation. Cooper (1965) comments that it is a 'better doer' than the form usually seen in cultivation. It agrees well with *P. diantha* described by Schlechter which Hunt & Vosa (1971) reduced to synonymy within *P. humilis*. However, a recent cytological study by Stergianou (1987) has given a chromosome count of 2n = 60 for this clone. She also reports that while two of the chromosome sets agree well with *P. humilis*, the other probably originates from *P. bulbocodioides*. This clone seems likely, therefore, to be a natural hybrid derived from the crossing of an unreduced gamete of *P. humilis* and a normal one of *P. bulbocodioides*.

Pleione humilis is less easily grown than many other species but is nevertheless relatively common in cultivation due, no doubt, to its frequent reintroduction from the wild. Grey-Wilson (pers. comm.) comments that it is a common species in eastern Nepal and characteristically grows epiphytically in moss, even on smooth-barked *Rhododendron* species, numerous plants forming rings or collars around the trunks and branches.

In cultivation *P. humilis* requires similar treatment to that of *P. hookeriana*. The pseudobulbs should be kept cold during the winter, 0–2°C being ideal, and the surface of the pan should be covered with live sphagnum moss. Care must be taken with the plants as they come into flower, for unlike other *Pleione* species, new roots are not produced at the base of the young shoots until later, indeed often not until the flowers have withered. As a result, the compost should only be kept very slightly moist until active root growth is under way—then more water can be applied. This species does not produce the usual two or three bulbils at the apex of the old pseudobulbs as in other species. Instead a mass of often fifty or more tiny bulbils appear. These are not very easy to grow on to maturity. The mature pseudobulbs do not always double themselves so that plants can be rather slow to increase.

Pleione humilis is most closely allied to the Chinese *P. albiflora*, which is likewise white-flowered with a crimson-blotched lip; it differs, however, in its smaller flowers and distinct callus.

Pleione humilis (J.E. Sm.) D. Don, Prodr. Fl. Nepal. 37 (1825); Hunt & Vosa in Kew Bull. 25: 428 (1971); Cribb et al. in Curtis's Bot. Mag. 184: 119, t.863 (1983).

Epidendrum humile J.E. Sm., Exot. Bot. 2: 75, t.98 (1806). Type: Nepal, *Buchanan-Hamilton* s.n. (holotype Hb. Linn.!).

Coelogyne humilis (J.E. Sm.) Lindl., Collect. Bot., sub. t.37 (1821).

C. humilis (J.E. Sm.) Lindl. var. *tricolor* Reichb. fil. in Gard. Chron., ser. 2, 13: 394 (1880). Type: origin unknown, cult. *Bull* (holotype W!).

C. humilis (J.E. Sm.) Lindl. var. *albata* Reichb. fil. in Gard. Chron., ser. 3, 3: 392 (1888). Type: origin unknown, cult. *Sander* (holotype W!).

Pleione humilis (J.E. Sm.) D. Don var. *adnata* Pfitz. in Engl., Pflanzenr. (IV.50) Coelog. 121 (1907). Type: not known.

P. humilis (J.E. Sm.) D. Don var. *purpurascens* Pfitz., op. cit. 122 (1907). Type: not known.

P. diantha Schltr. in Orchis 9: 44 (1915). Type: Burma, cult. *Hennis* (holotype B—destroyed).

P. humilis (J.E. Sm.) D. Don var. *pulchella* E.W. Cooper in Roy. Hort. Soc., Dict. Gard. 3: 1606 (1951). Type: not known.

Description. *An epiphytic or lithophytic herb. Pseudobulbs* flask-shaped with a long neck, 20–60 mm tall, 8–20 mm in diameter, olive-green, 1-leafed. *Leaf* oblanceolate to elliptic, 18–25 cm long, 2.8–3.5 cm wide, acute. *Inflorescence* with 1 or 2 flowers, produced before the leaf. *Peduncle* clothed in scarious sheaths. *Bract* obovate, subacute to obtuse, 20–32 mm long, 8–12 mm wide, subacute to obtuse. *Flowers* spreading to nodding; sepals and petals white; lip white spotted and streaked with crimson or yellow-brown with a central pale yellow zone. *Dorsal sepal* linear-oblanceolate, 34–47 mm long, 6–7 mm wide, subacute. *Lateral sepals* obliquely oblanceolate, 40–53 mm long, 70–90 mm wide, subacute. *Petals* obliquely linear-oblanceolate, 31–42 mm long, 5–7 mm wide, rounded at the apex. *Lip* oblong-elliptic, emarginate, saccate at the base, obscurely 3-lobed in front, 34–44 mm long, 25–31 mm wide, the margin lacerate in the apical half; side-lobes erect-incurved; callus consisting of 5–7 barbate lamellae. *Column* 26–28 mm long, broadly winged, irregularly dentate at the apex. *Ovary* 20–30 mm long. 2n = 40. PLATES 4, 15B, 20B; FIGURE 5A, p. 51.

DISTRIBUTION. Burma, NE India (Assam, Manipur & Sikkim) and Nepal; altitude 1850–3200 m. MAP 2, p. 38.

8. PLEIONE CORONARIA

Pleione coronaria is a remarkable species in several ways. It is the only Nepalese species with entire lamellae on the lip and in this character it appears to come closest to the Chinese *P. yunnanensis* and to the yellow-flowered *P. forrestii* from Burma and south China. This new species undoubtedly belongs in sect. *Pleione*, its single-leafed, conical-ovoid pseudobulbs and spring-flowering habit being characteristic of the other species of this section.

It was discovered by Adam Stainton, co-author with Oleg Polunin of *Wild Flowers of the Himalaya* (1984), in the Ganesh Himal of central Nepal in late April 1962. Stainton, in his diary, refers to woodland with many large *Juniperus recurva*, *Tsuga*, *Osmanthus suavis* and *Viburnum cordifolium* as the habitat of pink-flowered 'Pleione humilis'. This is undoubtedly *P. coronaria.*

The discovery of *P. coronaria* brings to five the number of species of *Pleione* recorded from Nepal. Of these, two others, *P. hookeriana* and *P. humilis*, are similarly spring-flowering whilst the others, *P. maculata* and *P. praecox*, both flower in the autumn. All of these are relatively widespread and locally abundant except for *P. coronaria* which is known only from the type, and one other collection from the Ganesh Himal to the north of Kathmandu.

Since this interesting species was first described, living material has been introduced into cultivation by Niklaus Trudel of Meilen, Switzerland. His account of his visit to the Ganesh Himal was published in the German Orchid Society's journal (*Die Orchidee* 36: 206–12), in 1985. The plants were collected by one of his Nepalese helpers, but Trudel suggests that the habitat was *Rhododendron* woodland at 3000–3500 m, with *P. coronaria* growing as an epiphyte in moss on branches and trunks down to ground level.

Pleione coronaria has already flowered in cultivation for Trudel and for one of the authors (IB), and the flowers agree well with those of the preserved type specimen. Trudel, however, notes that the pseudobulbs are ovoid rather than bottle-shaped when mature and have a less conspicuous crown at the apex than in *P. humilis* when the leaves fall. He has also noted that *P. coronaria* often bears two flowers in an inflorescence.

Pleione coronaria may well be endangered in its rather limited range for Trudel reports that the area has been heavily deforested and that the extensive reafforestation programme in Nepal has not yet reached the region.

With the precedent of both *P.* × *lagenaria* and *P.* × *confusa*, it is worth considering the possibility that *P. coronaria* may also have a hybrid origin. However, although it resembles *P. humilis* and *P. grandiflora* in vegetative characters, no cross between any of the well-known Nepalese species, neither spring- nor autumn-flowering, could possibly account for its flowers, and, in particular, the structure of its callus. The inevitable conclusion must be that *P. coronaria* is a distinct species of very restricted distribution.

Like *P. humilis*, this species also produces bulbils at the apex of the pseudobulbs. They can be successfully grown on by placing them in moist, but not wet, garden moss. In sphagnum moss they tend to rot.

Pleione coronaria Cribb & C.Z. Tang in Curtis's Bot. Mag. 184: 123 (1983). Type: Nepal; Ganesh Himal, *Stainton* 3654 (holotype BM!; icon. holotype K!).

DESCRIPTION. *An epiphytic herb. Pseudobulbs* clustered, often oblique, ovoid or ovoid-conical, up to 40 mm long and 18 mm in diameter, 1-leafed at the apex. *Leaf* spreading, lanceolate, acute, up to 15 cm long and 5 cm wide; leaf-base crown-like, 2–3 mm tall. *Inflorescence* erect, 1-flowered, 8–10 cm tall. *Bract* cucullate, elliptic, rounded, *c.* 27 mm long and 10 mm wide. *Flower* spreading to subnutant, pale mauvish pink, lip spotted with red or purple towards the apex and between the callus-lamellae. *Dorsal sepal* linear, 40–45 mm long, *c.* 6.5 mm wide, acute. *Lateral sepals* slightly falcate, lanceolate, 36–44 mm long, *c.* 9 mm wide, acute. *Petals* linear-oblanceolate, 40–53 mm long, *c.* 6 mm wide, obtuse. *Lip* somewhat oblong-elliptic, obscurely 3-lobed in the apical half, 40–49 mm long, 25–33 mm wide, united for the basal 5 mm to the side of the column, front margin denticulate; side-lobes obscure, erect-incurved; mid-lobe obflabellate, truncate, slightly emarginate; lamellae 6, possibly 7, entire, undulate, middle 4 longer than the outer ones, almost reaching the apex of the lip. *Column* slender, 31–32 mm long, erose at the apex. *Ovary* 15–21 mm long. *Fruit* ellipsoidal, 3 cm long. 2n = 40. PLATES 5, 16F; FIGURE 6E, p. 56.

DISTRIBUTION. C Nepal; altitude 2850–3500 m. MAP 2, p. 38.

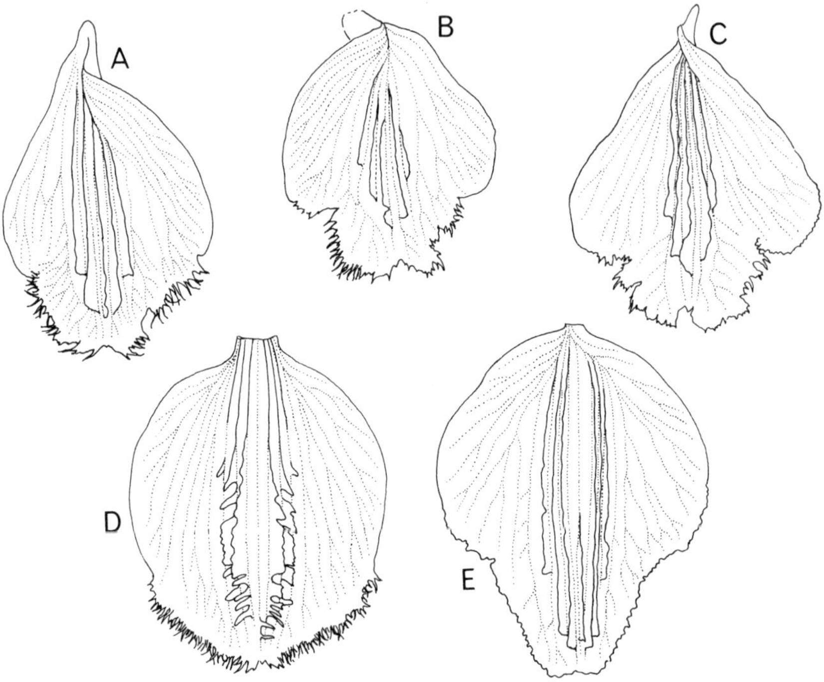

Figure 6. Lips of *Pleione*, ×1. A, *P. forrestii*, Forrest 16967; B, *P. forrestii*, Forrest 24425; C, *P. yunnanensis*, Henry 11113; D, *P.* × *confusa*, Kew Spirit Colln. 27197; E, *P. coronaria*, Stainton 3654.

9. PLEIONE FORRESTII

A great deal of confusion has resulted from the introduction into cultivation and successful propagation of plants reputed to be *P. forrestii*, but which have proved to be of hybrid origin (Harberd, 1980; Wimber & Cribb, 1981). Although the parentage of these plants cannot be proven, it seems they result largely from the hybridization of *P. forrestii* with the large white-flowered species *P. albiflora*, both of which grow together in the same habitat in adjacent parts of north Burma and west Yunnan.

True *P. forrestii* has recently been re-collected (Lancaster, 1982) by the 1979 Sino-British Expedition to Cangshan in Yunnan, and living plants (*SBEC* 0206), introduced into cultivation as a result, have already flowered at the Royal Botanic Garden, Edinburgh. The plants illustrated here were painted there in the spring of 1982.

Pleione forrestii is very distinctive in having yellow flowers and a lip-callus of five to seven entire lamellae. It is the only yellow-flowered *Pleione* and as such is a highly desirable plant, particularly as it has been used to introduce yellows into a succession of recent hybrids, notably *P.* Shantung. True *P. forrestii* differs from the hybrid widely grown under that name in having slightly smaller flowers with more falcate, acute petals and lateral sepals, and entire, rather than dissected, lamellae on the lip. The dissected lamellae betray the presence of *P. albiflora* in the hybrid.

Pleione forrestii, according to the field notes on the various herbarium collections, is variable in flower colour. The type is described as being bright orange with brown markings on the lip. The recent Lancaster importations are primrose-yellow with crimson spots on the lip. We have seen two recent (not Lancaster) introductions in flower and both are darker yellow than those collected by Lancaster but have paler lip markings; one of them had attractively scented flowers. In some other collections made by Farrer and later by Forrest the flowers are said to be creamy yellow or even ivory-white. It may be that the original introduction, which has subsequently proved to be a hybrid, was introduced from such colonies where the paler flowers may have resulted from the introgressive influence of the white-flowered *P. albiflora*. However, such ideas are mere conjecture until critical field studies have been made.

Pleione alba was described in 1984 by Li and Feng based on a collection by the latter from Dayao (Ta-yao) in west-central Yunnan in March 1982. This is only some 100 km east of the city of Tali (now Dali) in the vicinity of which *P. forrestii* is found. It seems surprising, therefore, that the authors compare *P. alba* with *P. yunnanensis* rather than with *P. forrestii*.

We have not seen the type of *P. alba*, but a good illustration accompanies the original description. This agrees well with drawings made from several *P. forrestii* collections we have seen. The only difference that is obvious is the flower colour of *P. alba* which is white with purple spots on the apical half of the lip. As noted by Cribb et al. (1983) the flower colour of *P. forrestii* varies from ivory-white through pale to deep yellow. The strong suggestion exists that *P. alba* is no more than a white-flowered variant of *P. forrestii* and we have, therefore, tentatively assigned it to the synonymy of the latter.

Pleione forrestii is, like *P. scopulorum* and *P. albiflora*, somewhat restricted in its distribution, having been found only in north Burma and west Yunnan. It grows there either in mats on mossy granitic boulders in somewhat shaded or exposed sites, on cliffs in deep shady valleys, or more rarely epiphytically on mossy tree-trunks.

Pleione forrestii Schltr. in Notes Roy. Bot. Gard. Edinb. 5: 106 (1912); Hunt & Vosa in Kew Bull. 25: 428 (1971); Lancaster in The Garden 107: 429 (1982); Cribb et al. in Curtis's Bot. Mag. 184: 123, t.864 (1983). Type: China; Yunnan, Tali [Dali] Range, *Forrest* 4859 (holotype E!).

? *P. alba* Li & Feng in Acta Bot. Yunn. 6(2): 193 (1984). Type: Yunnan, Dayao, *Feng* 82–77 (holotype KUN!).

DESCRIPTION. *A lithophytic or rarely an epiphytic herb. Pseudobulbs* conical, 15–30 mm long, 6–15 mm in diameter with a distinct collar-like leaf-base at the apex of the old pseudobulbs, 1-leafed. *Leaf* narrowly elliptic-lanceolate, up to 15 cm long, 1.5–4 cm wide, acute. *Inflorescence* 5–11 cm tall, appearing before the leaf, 1-flowered. *Peduncle* covered by 3 or 4 scarious, acute, sheathing bracts. *Bract* cucullate, oblanceolate, 24–35 mm long, 7–11 mm wide, subacute. *Flower* pale yellow to orange-yellow, rarely white, marked with brown or crimson spots on the lip. *Dorsal sepal* oblanceolate, acute, 30–40 mm long, 7–8 mm wide. *Lateral sepals* obliquely oblanceolate, 36–42 mm long, 7–8 wide, obtuse. *Petals* falcate, oblanceolate, subacute to acute, 36–42 mm long, 7–8 mm wide, subacute to acute. *Lip* elliptic-obovate in outline, 3-lobed, with lacerate margins in the apical half, 32–37 mm long, 27–31 mm wide; side-lobes erect, rounded in front; mid-lobe subquadrate, emarginate; callus consisting of 5–7 entire lamellae, the highest at the apex. *Column* arcuate, obscurely winged towards the apex, 26–32 mm long. *Ovary* 9–15 mm long. 2n = 40. PLATES 6, 16A; FIGURES 6A, 6B, p. 56.

DISTRIBUTION. N Burma and SW China (W Yunnan); altitude 2400–3100 m. MAP 3, p. 47.

10. PLEIONE × CONFUSA

Since its introduction from the wild by George Forrest, who sent pseudobulbs to J.C. Williams of Caerhays Castle in 1924, this plant has been grown under the name '*P. forrestii*'. However, recent morphological and cytological studies have indicated that it is in fact a plant of hybrid origin (Harberd, 1980; Wimber & Cribb, 1981).

The true *P. forrestii* Schltr. is certainly one of the parents of this hybrid, while *P. grandiflora* has been suggested as the other. To add to this confusion, however, the herbarium specimens from north-east Burma and west Yunnan previously referred to as *P. grandiflora* have been shown in the present account to belong to a recently described species, *P. albiflora* (see p. 49). Thus, it seems probable that *P.* × *confusa* is a natural hybrid between the sympatric species *P. forrestii* and *P. albiflora*. It may well be that its survival in cultivation is due to the relatively vigorous nature of the hybrid clone introduced by Forrest.

At one stage during the Second World War, this introduction was reduced to a single pseudobulb surviving at the Royal Botanic Garden, Edinburgh, but subsequently it responded well to improved cultivation by the Curator, E.E. Kemp, and a stock was slowly built up. Today this charming plant is far more widespread and, indeed, plants can be purchased through the horticultural trade, albeit at a rather inflated price. Peter F. Hunt gives a fuller account of its survival in cultivation, which accompanies a coloured illustration of the type plant, in *Curtis's Botanical Maazine* (t.501, 1967).

Pleione × *confusa* is slow to increase in cultivation. The pseudobulbs should be kept cold during the winter, a temperature just above freezing being ideal. The standard compost with extra sphagnum moss is most beneficial and, in addition, the pan should be surfaced with live sphagnum. During the summer months the leafy plants should be given plenty of fresh air, so a position near a ventilator or fan should be sought.

Pleione × *confusa* closely resembles *P. forrestii* in flower colour but differs in having larger flowers with longer, more falcate petals and a lip with cut, rather than entire, lamellae.

Pleione × confusa Cribb & C.Z. Tang in Curtis's Bot. Mag. 184: 126 (1983). Type: Kew Spirit Colln. no. 27197 (holotype K!).
P. forrestii sensu P.F. Hunt in Curtis's Bot. Mag. 176: t.501 (1967), non Schltr. (1912).

DESCRIPTION. *A terrestrial or lithophytic herb. Pseudobulbs* conical-ovoid, gradually narrowing into a beak, 20–25 mm long and wide when mature, obscurely angled when old, green or dull olive-green, 1-leafed at the apex. *Leaf* elliptic-lanceolate to oblanceolate, acuminate, 10–15 cm long, 3–4 cm wide, deciduous, grass-green. *Inflorescence* erect or suberect, 1-flowered. *Peduncle* 20–50 mm long, reddish. *Bract* oblanceolate, acute, 25 mm long, 10–15 mm wide, pale maroon with deeper venation. *Flower* primose-yellow to pale yellow, spotted with red on the lip, fragrant; pedicel and ovary 20 mm long, pale maroon. *Dorsal sepal* elliptic-lanceolate, acute, 38–42 mm long, 7–8 mm wide. *Lateral sepals* elliptic-oblanceolate, slightly oblique, acute, 37–40 mm long, 9–12 mm wide. *Petals* oblanceolate, acute or subacute, 38–42 mm long, 7–9 mm wide. *Lip* obscurely 3-lobed, 25–32 mm long, 30–35 mm wide, saccate at the base, emarginate at the apex, lacerate on margins in apical half; callus consisting of 4–6 raised ridges, undulate along their margin, the outer 2 short. *Column* clavate, slightly incurved and obscurely winged at the apex, 37–40 mm long. 2n = 40. PLATES 7, 16B; FIGURE 6D, p. 56.

DISTRIBUTION. SW China (Yunnan) and possibly adjacent N Burma, where the parents are found growing near one another.

NOTE. A specimen sent from the Royal Botanic Garden, Edinburgh was used as the basis for the plate in *Curtis's Botanical Magazine* (t.501, 1967). This same specimen was placed in spirit and presents the type collection of *P.* × *confusa.*

11. PLEIONE YUNNANENSIS

The history of the name *P. yunnanensis* illustrates well the problems caused by the misidentification of plants introduced into cultivation. This species was described in 1903 by Rolfe based on a Henry collection (no. 11113) from Mengtze in southern Yunnan. Some three years later, plants which had been grown by Messrs Sutton of Reading, who received them from China, were figured in *Curtis's Botanical Magazine* (t.8106) as *P. yunnanensis*. The plate was accompanied by a brief account by Rolfe who had identified them as his recently described species. However, a careful comparison of the type specimen of *P. yunnanensis* with the illustration shows that Sutton's plants differed in having much larger flowers, an ovary consistently shorter than the subtending bract, more acute sepals and petals, a longer column and, most noteworthy, a callus of ragged lamellae. In every way these plants agree with *P. bulbocodioides*, not with *P. yunnanensis*. Sutton's plants have persisted in cultivation until the present day with the name that Rolfe mistakenly attached to them. Hunt & Vosa (1971) did not realize the error and hence their chromosome count of 2n = 120 for *P. yunnanensis* must now be attributed to *P. bulbocodioides*.

Rolfe's mistake finally came to light when plants of true *P. yunnanensis* were recently introduced into cultivation from Yunnan from several separate sources. They proved quite distinct from the plants illustrated in *Curtis's Botanical Magazine* when they flowered, and when compared with herbarium material were found to match well the type of *P. yunnanensis*.

Pleione yunnanensis is one of the smaller-flowered species with attractive lavender-pink to deeper rose-coloured flowers with purple spots and streaks on the lip. An albino variant has recently been flowered by H. H. Pinkepank as *P. yunnanensis* 'Yeti'. It differs from allied species such as *P. bulbocodioides* in having smaller, paler coloured flowers, rounded petals, a bract which is shorter than the pedicel and ovary, and a lip which has broad rounded side-lobes, a subquadrate dentate mid-lobe and five entire white lamellae.

Pleione yunnanensis is found in the northern extremity of Burma, in neighbouring Yunnan and as far north as south Sichuan and west Guizhou. It is found growing in grassy meadows or on rocks in shady places. An illustrated account of its recent collection in Yunnan and details of its habitat are given by Lancaster (1982). One of us (PC) has recently seen it growing in the mountains at 2600 m between Chuxiong and Xiaguan in west Yunnan. It grows there in well-spaced colonies in open *Pinus yunnanensis* forest under Ericaceous shrubs, notably *Rhododendron* and *Lyonia*, amongst rocks and in deep leaf-litter with the pseudobulbs well buried, 2–5 cm below ground level. In some places it must grow with *P. bulbocodioides* as *Delavay* 3347a is in fact a mixed collection of the two species.

It seems likely that *P. yunnanensis* hybridizes occasionally in the wild with *P. bulbocodioides* and that subsequent backcrossing to both parents has led in some localities to the production of hybrid swarms (Harberd, pers. comm.; Grey-Wilson, pers. comm.), PLATE 19C. We have seen such putative hybrids and backcrosses in several recent introductions, only one of whose provenance is known, a collection from the Cang Shan mountains above the city of Dali in western Yunnan.

Detailed accounts of the cytology of both species and hybrids are given by Wimber & Cribb (1981) and Stergianou (1987).

Pleione yunnanensis (Rolfe) Rolfe in Orch. Rev. 11: 292 (1903); Hunt & Vosa in Kew Bull. 25: 428 (1971); Lancaster in The Garden 107: 428 (1982); Cribb et al. in Curtis's Bot. Mag. 184: 127, t.865 (1983).

Coelogyne yunnanensis Rolfe in Journ. Linn. Soc. 36: 23 (Jan. 1903). Type: China, Yunnan, Mengtze, *Henry* 11113 (lectotype K!).

P. chiwuana Tang & Wang in Acta Phytotax. Sin. 1: 78 (1951). Type: Yunnan, *Wang* 62698 (isotype K!).

DESCRIPTION. *A terrestrial or lithophytic herb. Pseudobulbs* conical, 10–20 mm tall, 10–15 mm in diameter, 1-leafed. *Leaf* erect, lanceolate to narrowly elliptic, 15–25 cm long, 2.8–6 cm wide, acuminate; petiole 1.5–5.5 cm long. *Inflorescence* erect, 1-(rarely 2-)flowered, appearing before the leaves. *Peduncle* 7–15 cm long. *Bract* obovate, truncate, mucronate, 2–2.8 cm long, shorter than the pedicel and ovary. *Flower* spreading, pale lavender to rose-pink or rarely white, spotted on the lip with purple or red; pedicel and ovary ± deep red. *Dorsal sepal* oblong-oblanceolate, 33–40 mm long, 6–7 mm wide, obtuse. *Lateral sepals* obliquely oblong- to elliptic-lanceolate, 33–40 mm long, 6–10 mm wide, obtuse. *Petals* oblanceolate, 30–42 mm long, 5–10 mm wide, rounded at the apices. *Lip* 3-lobed, 30–38 mm long, 23–30 mm wide, with a dentate apical margin; side-lobes erect, broad, rounded in front; mid-lobe subquadrate, emarginate and mucronate at the apex; callus consisting of 5 entire lamellae. *Column* broadly winged, 18–22 mm long. *Ovary* 26–42 mm long. *Fruit* fusiform-cylindric, 25–35 mm long. 2n = 40. PLATE 8; FIGURE 6C, p. 56.

DISTRIBUTION. N Burma and China (Yunnan, Sichuan & Guizhou); altitude 1350–3200 m. MAP 3, p. 47.

12. PLEIONE GRANDIFLORA

Pleione grandiflora is known only from the type collection from Mengtze in south Yunnan. A second specimen (*Henry* 11115), also from the Mengtze region and said to have pink flowers, resembles the type in its habit but has smaller bracts and lacks flowers so that its true identity must remain speculative.

Pleione grandiflora is allied to *P. yunnanensis* but differs in its much larger pseudobulbs, and larger differently coloured flowers with lacerate rather than entire lamellae. The lamellae are more reminiscent of those of *P. bulbocodioides* Rolfe which, however, has smaller distinctly coloured flowers each with a shorter broader bract, and a far smaller pseudobulb.

The type collection comes from montane forest at an altitude of about 2700 m. The introduction of this large-flowered species into cultivation is eagerly anticipated. Earlier suggestions that it was one parent of the hybrid mistakenly grown for many years as *P. forrestii* are undoubtedly incorrect, the confusion having arisen from the misidentification of specimens of *P. albiflora* as *P. grandiflora*.

Pleione grandiflora (Rolfe) Rolfe in Orch. Rev. 11: 291 (Oct. 1903); Hunt & Vosa in Kew Bull. 25: 428 (1971); Cribb et al. in Curtis's Bot. Mag. 184: 129 (1983).

Coelogyne grandiflora Rolfe in Journ. Linn. Soc. 36: 22 (Jan. 1903).
Type: China, Yunnan, Mengtze, *Henry* 11116 (holotype K!).

DESCRIPTION. *A lithophytic herb. Pseudobulbs* obliquely conical, 30–45(–70) mm tall, 12–15 mm in diameter, surrounded by fibrous remains of basal sheaths, 1-leafed. *Leaf* lanceolate, acute, not fully developed in material seen. *Inflorescence* erect, 1-flowered. *Peduncle* 8–12 cm long, covered in the basal two-thirds by 3 or 4 tubular, subacute sheaths. *Bract* cucullate, narrowly elliptic, 23–40 mm long, obtuse. *Flowers* white or possibly sometimes pink. *Dorsal sepal* oblanceolate, 50–55 mm long, *c.* 10 mm wide, rounded at the apex. *Lateral sepals* oblique, narrowly elliptic, *c.* 50 mm long and 12 mm wide, rounded at the apices. *Petals* falcate, oblanceolate, *c.* 55 mm long and 8.5 mm wide, rounded at the apices. *Lip* obscurely 3-lobed, emarginate, flabellate, 50–55 mm long, *c.* 36 mm broad, margin in apical half coarsely lacerate: callus consisting of 5 lacerate lamellae. *Column* arcuate, very narrowly winged at the apex, entire on apical margin, 38–40 mm long. *Ovary c.* 15 mm long. FIGURES 5B, 7, pp. 51 and below.

DISTRIBUTION. SW China (S Yunnan); altitude 2650–2850 m. MAP 3, p. 47.

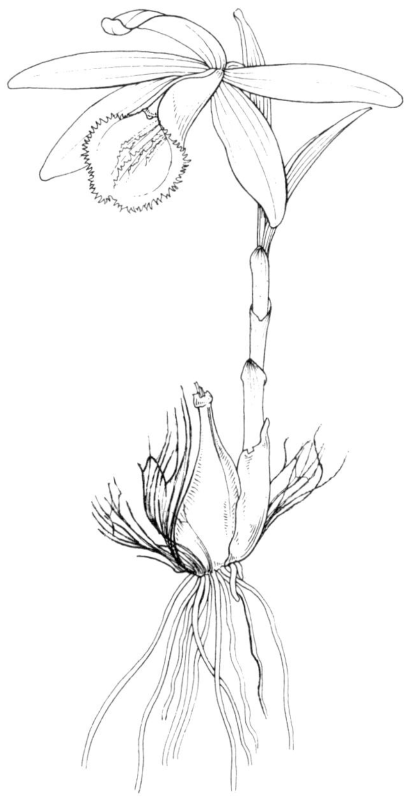

Figure 7. *Pleione grandiflora*, ×¾, Henry 11116.

THE PLEIONE BULBOCODIOIDES COMPLEX
(Species 13–16)

The following species—*P. bulbocodioides*, *P. formosana*, *P. limprichtii*, *P. speciosa* and the little-known *P. chunii*—are all closely allied and have been recently considered, by at least some authorities, to represent one, albeit rather variable, species—*P. bulbocodioides* (see Hunt & Vosa, 1971). However, there are in cultivation, in Europe at least, four entities which although obviously allied are yet, in the living state, relatively easy to distinguish.

Members of this complex flower in the spring, either before the leaves or as the leaves develop, and can be recognized by their predominantly pink to magenta, less commonly white, flowers. The lip bears two to five, more or less erose, denticulate or lacerate lamellae.

Following a request from the Royal Horticultural Society to clarify the confused nomenclature of the genus, Hunt & Vosa (1971) concluded that all of this complex should be treated as one variable species, *P. bulbocodioides*, which is the earliest applicable name. However, this treatment has not been at all widely followed and the Royal Horticultural Society continues to register hybrids between species considered conspecific by Hunt & Vosa. A reassessment is then obviously needed, particularly in the light of the long-standing misidentification of the plants grown for many years under the names *P. yunnanensis* and *P. pogonioides*!

Several factors have confounded attempts to produce a sensible account of the 'bulbocodioides complex' as it will henceforth be referred to in this account. First, its widespread distribution from Xizang (Tibet) in the west to Taiwan in the east has meant that some 14 or 15 different names have been applied to elements of this complex in the past. New species have often been described by botanists who have not had access to the type material of species already described. This is particularly distressing in the case of both Schlechter and Kraenzlin names, for their types were destroyed in Berlin and only a few isotypes remain extant.

Although there have been a large number of introductions from Taiwan, only very few clones have been introduced into cultivation from mainland China. These clones do indeed look different from the Taiwanese material and may represent distinct species. However, the possibility that they may represent selected forms from a range of more or less continuous variation running from Xizang eastwards to

Plate 11

Pleione limprichitii (top)

Pleione speciosa (bottom)

CHRISTABEL KING

Plate 12

Pleione Alishan 'Merlin' (top)
Pleione Stromboli 'Fireball' (middle)
Pleione Brigadoon (bottom)

CHRISTABEL KING

Plate 13

Pleione Shantung (top)

Pleione Vesuvius 'Phoenix' (middle)

Pleione Soufrière (bottom)

CHRISTABEL KING

Plate 14

Pleione Versailles 'Swallow' (top)
Pleione Eiger (middle)
Pleione Hekla (bottom)

CHRISTABEL KING

Taiwan cannot be ignored. Although a considerable number of Chinese collections are represented in the world's herbaria, the difficulty of handling pressed material cannot compensate for the lack of authenticated living material.

The same applies also to the final factor which affects the group, which is polyploidy. Hunt & Vosa (1971) and Wimber & Cribb (1981) have demonstrated the presence of diploids, tetraploids and even a hexaploid clone in the limited sample of living plants examined.

The following treatment represents a reassessment of the complex based on relatively extensive floral dissections from material loaned by the Herbaria at Kew, Paris, Edinburgh, the British Museum, Kunming and Chengdu, and living material in the collections of one of the authors (IB) and of the Royal Botanic Gardens, Kew.

In the last two or three years Chinese material assignable to this complex has been appearing in the nursery trade in increasing quantities. However, few of these introductions are from known sources and they have done little to clarify the confused taxonomy of this group.

13. PLEIONE BULBOCODIOIDES

Pleione bulbocodioides is closely allied to *P. speciosa*, the two being apparently sympatric in central China. However, it differs in having smaller flowers with a shorter column and a distinctly straighter lip which is flared at the apex and usually with four or five rather irregularly lacerate lamellae. This species is also allied to *P. formosana* but differs in having smaller, more evenly coloured, darker purple flowers with lacerate lamellae on the lip.

Plants introduced from China by Messrs Sutton of Reading and figured in 1906 in *Curtis's Botanical Magazine* (t.8106) were misidentified by Rolfe as *P. yunnanensis*, and under this name they have survived in cultivation until the present day. They can be attributed to *P. bulbocodioides* and are similar to plants recently introduced from Yunnan under the synonym *P. delavayi* (Lancaster, 1982). Yet, Hunt & Vosa (1971) did not appreciate this anomaly and continued to call Sutton's plants *P. yunnanensis*, an altogether distinct species.

Added to this confusion is the fact that both *P. pogonioides* and *P. speciosa* have also been misidentified, so it would seem best to re-examine the whole complex and to examine as many types as possible to resolve the nomenclatural muddle.

Pleione bulbocodioides appears to be a widespread and rather variable species. Three flowers of the type collection from east Xizang (Tibet) have been examined and have shown variability in the flower size and in the structure of the callus. The types of *P. henryi*, *P. pogonioides* and *P. communis* var. *subobtusum* do not differ significantly from that of *P. bulbocodioides*. We have also examined the types of *P. delavayi* and *P. communis* from Yunnan which differ in having a rather taller inflorescence, slightly larger flowers and a lip with an additional short but high lamella on the mid-lobe. Although this, at first, appears sufficient to recognize *P. delavayi* (the earlier name) as distinct there are several collections from north-west Yunnan and south Sichuan of an intermediate nature, with a much reduced mid-lamella and a shorter inflorescence.

No type material of *P. mairei*, *P. smithii* and *P. rhombilabia* nor of *P. mandarinorum*, has been seen but all appear from the type descriptions to fall within the range of variability of *P. bulbocodioides*. One of us (PC) has seen many plants of *P. bulbocodioides* in the wild in the area between Dali and Lijiang in north-west Yunnan. It is common throughout the region growing on steep banks and ridges in deep leaf-litter usually under *Rhododendron* bushes in open *Pinus* and *Tsuga* forest. It is very variable in flower colour (ranging from pale pink to deep rose-purple), flower size, peduncle length and pseudobulb size and shape.

Some of this variability may be explicable in terms of polyploidy, for of the two clones counted, one has proved to be diploid and the other hexaploid (Wimber & Cribb, 1981). Obviously more information is required before patterns of variability can be related to chromosome number, but new introductions are beginning to appear in cultivation and these may allow a more sensible assessment of specific limits in this complex to be made. It may well be, as Hunt & Vosa have suggested, that the variability is best expressed by recognizing sub- or infra-specific taxa within *P. bulbocodioides*.

Of the clones of *P. bulbocodioides* in cultivation one of the most striking is 'Chairman Mao' with a deep-coloured flower that is produced late in the spring.

Pleione bulbocodioides has the reputation of being difficult to flower. However, regular feeding will help to ensure large flowering-sized pseudobulbs. The pseudobulbs thrive best when buried in the compost, although even under perfect conditions this is a slow species to increase.

Pleione bulbocodioides (Franch.) Rolfe in Orch. Rev. 11: 291 (1903); Hunt & Vosa in Kew Bull. 25: 427 (1971) pro parte; Cribb et al. in Curtis's Bot. Mag. 184: 132, t.866 (1983).

Coelogyne bulbocodioides Franch. in Nouv. Arch. Mus. Paris, sér. 2, 10: 84 (1888). Type: Tibet, Moupine, *David* s.n. (holotype P!, isotype K!).

C. delavayi Rolfe in Bull. Misc. Inf. Kew 1896: 195 (1896). Type: China, Yunnan, *Delavay* 4739 (holotype K!).

C. henryi Rolfe, loc. cit. (1896). Type: China, Hupeh, *Henry* 6088a (lectotype K!).

C. pogonioides Rolfe, op. cit. 196 (1896). Type: China, Hupeh, Patung, *Henry* 3785 (lectotype K!).

Pleione delavayi (Rolfe) Rolfe in Orch. Rev. 11: 291 (1903).

P. mairei Schltr. in Feddes Rep. Sp. Nov. Beih. 4: 61 (1919). Type: China, Yunnan, *Maire* 6341 (not seen).

P. henryi (Rolfe) Schltr., op. cit. 186 (1919).

P. smithii Schltr. in Act. Hort. Goth. 1: 149 (1924). Type: China, Szechuan, Teng-hsiang-ying, *H. Smith*, 1883 (holotype GB!).

P. communis Gagnep. in Bull. Soc. Bot. France 78: 25 (1931). Types: China, Yunnan, *Ducloux* 2057, 5645, *Delavay* 3141; Tibet, *Monbeig* s.n. (syntypes P!).

P. communis Gagnep. var. *subobtusum* Gagnep., loc. cit. (1931). Types: Tibet, Gnia-pa-tong, *Soulié* 1385 & Tse-Kou, *Soulié* s.n. (syntypes P!).

P. ganchuenensis Gagnep., op. cit. 26 (1931). Type: China, Kouytcheou, Gan-chouen, *Cavalerie* s.n. (holotype P!).

P. fargesii Gagnep., op. cit. 25 (1931). Type: China, Tchen-Keou-Tin, *Farges* 534 (isotype K!).

P. rhombilabia Hand.-Mazz., Symb. Sin. 7: 1348 (1936). Type: China, Yunnan, nr. Lidjiang (=Lijiang), '*v. E.*' 3989 (not seen).

P. yunnanensis sensu Rolfe in Curtis's Bot. Mag. 176: t.8106 (1967), non Rolfe (1903).

Description. *A terrestrial or lithophytic herb. Pseudobulbs* conical to pyriform, distinctly necked, 20–26 mm long, 12–20 mm in diameter, 1-leafed. *Leaf* developing after the inflorescence, narrowly elliptic-lanceolate, 14 cm long or more, 2.5 cm wide or more, acute. *Inflorescence* erect, up to 20 cm long, 1-flowered. *Peduncle* 6–15.5 cm long, covered by 3 tubular sheaths in the basal half. *Bract* elliptic-obovate, 29–40 mm long, rounded or obtuse. *Flowers* pink to rose-purple or magenta with darker purple markings on the lip. *Dorsal sepal* oblanceolate, 34–45 mm long, 5–8 mm wide, obtuse. *Lateral sepals* oblique, narrowly elliptic, 33–44 mm long, 6–10 mm wide, subacute or obtuse. *Petals* oblique, oblanceolate, 37–46 mm long, 4–7 mm wide, acute or subacute. *Lip* obovate in outline, obscurely 3-lobed, 32–45 mm long, 25–35 mm wide, with lacerate margins in the apical half; mid-lobe subquadrate, emarginate; callus consisting of 4 or 5 erose lamellae, the middle one, if present, shorter but higher than the others. *Column* somewhat arcuate, 27–36 mm long. 2n = 40, 120. PLATES 9, 16C, 16D, 16E, 20A; FIGURES 8D, 8E, opposite.

DISTRIBUTION. China (Anhui, Gansu, Guizhou, Hubei, Sichuan, SE Xizang (Tibet) & Yunnan); altitude 900–3600 m. MAP 4, below.

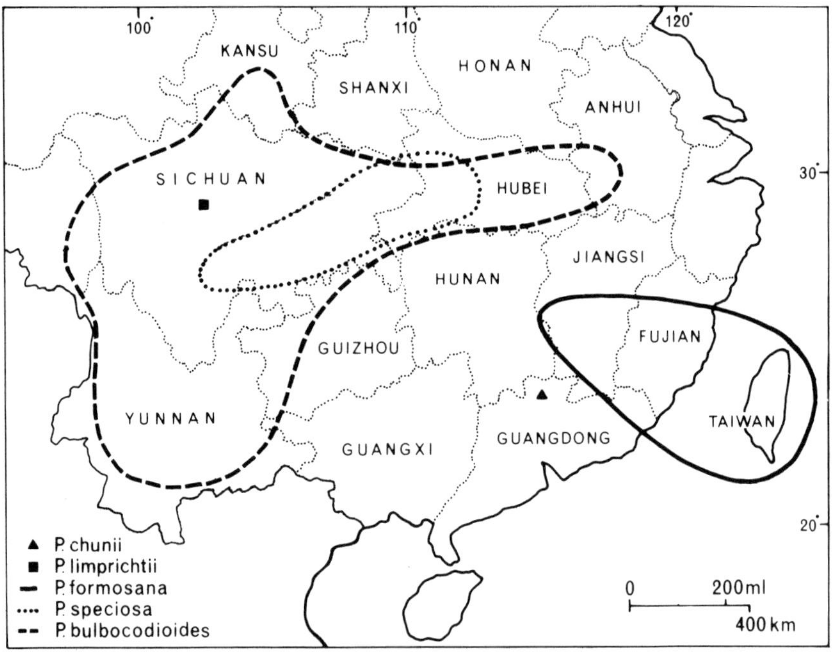

Map 4. Distribution of *Pleione* species.

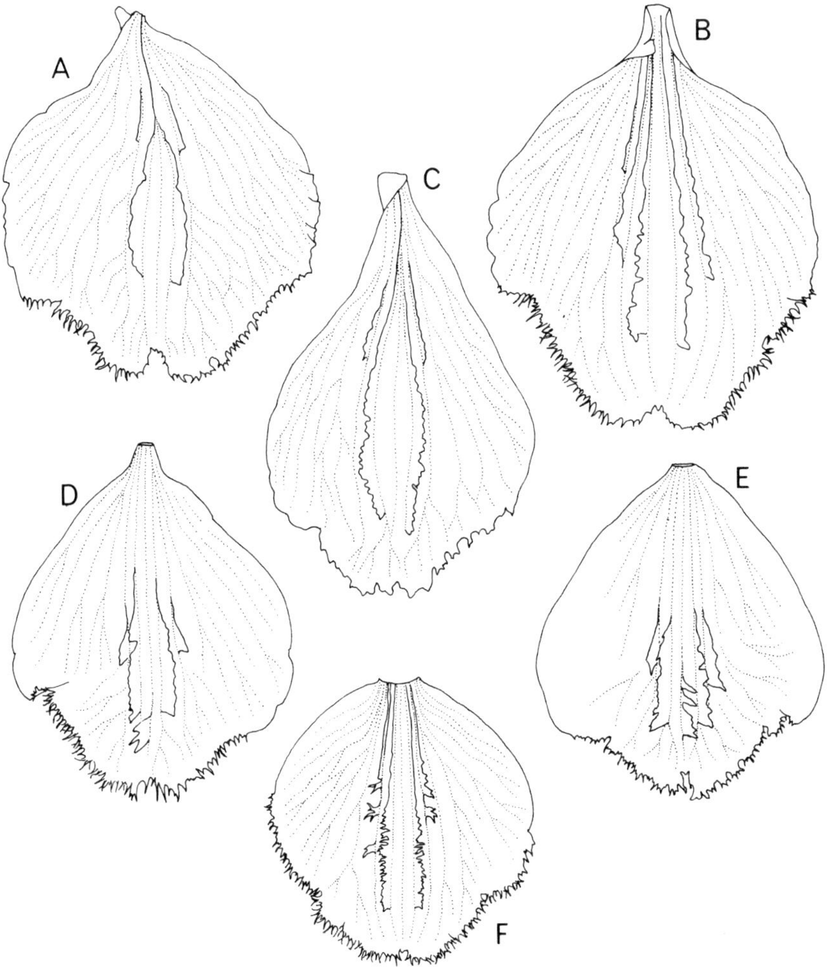

Figure 8. Lips of *Pleione*, ×1. A, *P. formosana*, Price s.n.; B, *P. formosana*, Rawinsky s.n.; C, *P. speciosa*, Wilson 1761; D, *P. bulbocodioides*, RBG Edinburgh ex Crossland; E, *P. bulbocodioides*, RBG Edinburgh acc. 771660; F, *P. limprichtii*, Kew Spirit Colln. 43997.

14. PLEIONE FORMOSANA

Pleione formosana is by far the commonest species of the genus in cultivation and also one of the hardiest. Some clones will survive quite happily when planted out-of-doors on a peat bed in a sheltered place in milder districts and will survive quite severe frosts in such situations.

This species is closely allied to both *P. bulbocodioides* and *P. speciosa*, differing from the former in its flower colour, larger flowers and callus in which the keels are often interrupted and usually entire, or not markedly erose, on their margins. From the latter, it differs again in flower colour and callus structure and also in having a shorter column and a lip which flares along the apical margin.

Pleione formosana, as known in cultivation, is an extremely variable species particularly in flower colour and size and in the number of lamellae on the lip. Several of the more attractive clones have received names and a list of these is given below. Considerable controversy has surrounded the status of *P. pricei*, living plants of which were introduced into the British Isles by William R. Price from Taiwan in 1912. Rolfe described and named it shortly afterwards and it has survived in cultivation under that name until the present day. The type differs from that of *P. formosana* in having two rather than four lamellae on the lip and a shorter one-flowered scape. However, many plants intermediate in character have been introduced into cultivation in recent years and it is generally agreed that *P. pricei* is conspecific with *P. formosana*, warranting only cultivar status as *P. formosana* 'Oriental Grace' and 'Oriental Splendour'.

Although all the cultivated material of *P. formosana* originates from Taiwan, its distribution almost certainly extends to mainland China. Examination of an isotype of *P. hui* Schltr. from Jiangsi Province indicates that it is probably conspecific with *P. formosana*, while another collection from Fujien Province also falls within its range of variation. Unfortunately, no material from these provinces of mainland China is currently in cultivation outside China.

Pleione formosana Hay. in Journ. Coll. Sci. Tokyo 30, art. 1: 326 (1911); Hunt in Curtis's Bot. Mag. 174: t.421 (1962); Cribb et al. in Curtis's Bot. Mag. 184: 135 (1983). Type: Taiwan; Nanõ, Hyanõsha, *Mori* 16 (holotype probably TI).

P. pricei Rolfe in Curtis's Bot. Mag. 143: t.8729 (1917). Type: Taiwan, Bonbonsan, Giran, *Price* s.n. (holotype K!).

P. hui Schltr. in Feddes Rep. Sp. Nov. 19: 377 (1924). Type: China, Kiangsi, *Hu*, 718 (isotype K!).

DESCRIPTION. *A terrestrial, lithophytic or epiphytic herb. Pseudobulbs* compressed-ovoid to ovoid, 13–30 mm long, 17–37 mm in diameter, green to dull dark purple, 1-leafed at the apex. *Leaf* elliptic to oblanceolate, 10–25 cm long, 3–5 cm wide, acute to obtuse, petiolate. *Inflorescence* up to 25 cm tall, 1- or less commonly 2-flowered. *Peduncle* 7–12 cm long. *Bract* linear-lanceolate to elliptic or obovate, 24–43 mm long, acute. *Flower* white to rose-pink, with yellow, reddish or brown marks on the lip and with a yellow base, sometimes weakly fragrant. *Dorsal sepal* narrowly elliptic-oblanceolate, 42–57 mm long, 9–15 mm wide, acute. *Lateral sepals* oblique, narrowly elliptic, 40–55 mm long, 10–15 mm wide, acute or subacute. *Petals* linear-oblanceolate, 42–60 mm long, 10–15 mm wide, subacute. *Lip* obscurely 3-lobed, 40–55 mm long, 32–46 mm wide, emarginate, with lacerate margins in the apical half; mid-lobe obscurely subquadrate; callus consisting of 2–5 interrupted lamellae with ± entire or erose margins, the central one, if present, very short. *Column* 28–36 mm long, obscurely dentate at the apex. 2n = 40. FIGURES 8A, 8B, p. 69.

DISTRIBUTION. E China (Fujien, Jiangsi & possibly Guangdong) and Taiwan. MAP 4, p. 68.

CULTIVARS. All the cultivars of *P. formosana* are very easy to grow and flower. A selection of the best is given here:

'Achievement'. Pseudobulbs purplish; flowers pale mauve-pink with a large open lip heavily blotched with red-brown and yellow.

'Ben Nevis'. Pseudobulbs greenish black; flowers large with pale lilac reflexed sepals and a lip with a wide pink margin with reddish brown markings.

'Blush of Dawn'. PLATE 10. Pseudobulbs green; flowers large, creamy white flushed with pale pink, lip marked with yellow and brown.

'Cairngorm'. Pseudobulbs green; flowers white, lip with brick-red markings and pale yellow lamellae.

'Clare'. Pseudobulbs very large, green; flowers often 2 per stem, large, white, lip marked with pale lemon-yellow.

'Crinoline'. Pseudobulbs flattish, dark green; flowers borne on very tall stems, fragrant, pale lilac-pink with a white lip marked with a few pale lilac spots.

'Emperor'. Pseudobulbs blackish green; flowers pale mauve-pink, lip frilled, white, lightly flushed with yellow and marked with red-brown within.

'Goldilocks'. Pseudobulbs blackish green; flowers mauve-pink with a paler lip heavily flushed with dark yellow and with large brown spots within.

'Iris'. Pseudobulbs dark purplish; flowers large, pale rose-purple, lip frilled, marked with red-brown and yellow.

'Lilac Beauty'. Pseudobulbs dark olive-green; flowers earlier than most other cultivars, lilac-pink with a paler lip heavily spotted with deeper lilac-pink.

'Lindsay Forbes'. Pseudobulbs greenish black; flowers large, mauvish pink with a paler lip which has a wide pink margin, marked with yellow and reddish brown.

'Oriental Grace'—(originally *P. pricei*). PLATE 10. Pseudobulbs purplish black; flowers with narrow rosy violet sepals and petals and a paler lip marked with yellow and brown.

'Oriental Jewel'. Pseudobulbs small, blackish purple; flowers later than most other cultivars, small, pale rose-purple, lip almost white marked with brownish yellow.

'Oriental Legend'. Pseudobulbs dark blackish green; flowers usually 2 per stem, dark rose-purple, lip frilled with a broad pink margin, marked yellow within.

'Oriental Splendour'—(originally *P. pricei*). Pseudobulbs purplish black; flowers usually 2 per stem, very freely produced, deep rose-violet, lip much paler, marked with brown and yellow.

'Orwell Glory'. Pseudobulbs large, blackish green; flowers scented, large, pink.

'Pitlochry'. Pseudobulbs large, olive-green; flowers very pale, almost white, lip marked with pale brown spots within.

'Polar Sun'. PLATE 10. Pseudobulbs small, green; flowers small, white, lip marked with yellow.

'Purple Emperor'. Pseudobulbs blackish purple; flowers deep rose-purple, lip white flushed with rose-purple and marked with darker rose-purple.

'Serenity'. Pseudobulbs large and squat, wrinkled, blackish purple; flowers very large, pale rose-purple, lip almost white marked with orange-brown.

'Snow White'. PLATE 19A. Pseudobulbs green; flowers appearing very early, often 2 per stem, glistening white, lip finely frilled, with lemon-yellow markings.

15. PLEIONE LIMPRICHTII

This is another of the species considered by Hunt & Vosa (1971) to be synonymous with *P. bulbocodioides*. However, the limited material we have observed differs from other species in the complex in having relatively smaller flowers, a short column and an almost circular lip (when flattened) with four regular denticulate or erose lamellae on the upper surface.

There are only a few (possibly no more than three) clones in cultivation, one of which is distinguished by growers as 'Pink' *limprichtii*. Hunt (1962) and Johnson (1986) have documented the introduction of the plants cultivated as *P. limprichtii*. These can be traced to material collected by the Swedish botanist Dr Harry Smith, near Kang-ting in Sichuan Province in south-west China in 1934 and sent to the nursery of Magnus Johnson at Södertälje for multiplication. The plant figured in *Curtis's Botanical Magazine* (t.397, 1962) came from this source via the nursery of Van Tubergen. Smith made two collections—no. 13018 from mountains east of Kang-ting at 2900–3050 m growing on cliffs and the other, no. 13020, at the same altitude on an outcrop of limestone just outside the east gate of the city. One of Limpricht's original collections (no. 1598) came from nearby, *c.* 30 km east of Kang-ting. Limpricht noted *Clintonia udensis, Geranium pylzowianum, Primula muscarioides* and *P. veitchii* all growing nearby.

In the wild *P. limprichtii* grows in colonies in a thin humus-rich soil sparsely covered by moss. In its natural habitat, the winters can be severe and the pseudobulbs may well be covered by snow for periods. It is scarcely surprising, therefore, that *P. limprichtii* has proved to be the hardiest species of the genus in cultivation. In dry conditions, Johnson reports that mature pseudobulbs can survive −20°C. Both of Smith's collections survive to the present day in cultivation and some variation is present in pseudobulb coloration and flower production but the flowers are recognizably of the same species.

In contrast, it seems likely that the specimens cultivated as 'Pink' *limprichtii* originated in Burma and were introduced by Frank Kingdon Ward. Certainly material of 'Pink' *limprichtii* presently in cultivation agrees well with a preserved flower of *Kingdon Ward* 13033 which flowered at Kew in 1957. The provenance of this specimen is given as Mt. Victoria in Burma, a considerable distance away from the Sichuan locality.

Unfortunately, despite examination of numerous herbarium specimens, we have been unable to find any intermediate localities.

Cytological studies by Hunt & Vosa (1971) and Wimber & Cribb (1981) have reported both diploid and tetraploid counts for *P. limprichtii*, confirming that two distinct clones (at least) exist in cultivation.

Pleione limprichtii is one of the most delightful species in cultivation, though not the easiest to satisfy. The plants should be kept cold during the winter months. Plants respond best when the pseudobulbs are buried in the compost, rather than resting half out of it.

Pleione limprichtii Schltr. in Feddes Repert. Beih. 12: 346 (1922); Hand.-Mazz., Symbol. Sin. 7: 1348 (1936); Adey in Gard. Chron. 145: 296 (1959); Hunt in Curtis's Bot. Mag. 174: t.397; (1962); Cribb et al. in Curtis's Bot. Mag. 184: 138, t.868 (1983); Johnson in Kew Mag. 3: 73 (1986). Type: China; Sichuan, *Limpricht* 1598 (isosyntype K!).

DESCRIPTION. *A terrestrial or lithophytic herb. Pseudobulbs* conical-ovoid, sometimes more elongate, 30–40 mm long, 20–25 mm in diameter, pale or deep green to purple, 1-leafed. *Leaf* appearing after the inflorescence, lanceolate, up to 13 cm long and 4 cm wide, acute. *Inflorescence* erect, 1- or rarely 2-flowered. *Peduncle* up to 10 cm long. *Bract* oblanceolate, 22–25 mm long, 6–8 mm wide, pale magenta, acute. *Flower* 40–55 mm across, pink to rose-magenta, lip spotted with brick-red and with white lamellae. *Dorsal sepal* narrowly elliptic, 30–35 mm long, 5–9 mm wide, acute. *Lateral sepals* oblique, narrowly elliptic, 28–35 mm long, 6–10 mm wide, acute. *Petals* falcate, oblanceolate, 30–35 mm long, 4–5 mm wide, acute. *Lip* ± orbicular, obscurely 3-lobed towards the apex, emarginate, 25–40 mm long, 25–35 mm wide, with lacerate margins in the apical half; callus consisting of 4 regularly denticulate or erose lamellae. *Column* slightly incurved, 25–30 mm long, winged towards the apex. 2n = 40, 80. PLATE 11; FIGURE 8F, p. 69.

DISTRIBUTION. SW China (Sichuan) and possibly N Burma; altitude *c.* 2000–2500 m. MAP 4, p. 68.

16. PLEIONE SPECIOSA

Pleione speciosa is probably the most strikingly attractive species in the 'bulbocodioides complex' with its large rich rose-purple or magenta flowers with yellow lamellae on the lip. It is similar in flower size to *P. formosana* but differs in its richer flower colour and in having markedly falcate lateral sepals. Furthermore, the almost entire obovate or subrhombic lip is noticeably hooked in the apical half (viewed sideways) and does not have a flared apical margin. The callus consists of two or four erose or denticulate lamellae and the column is longer than that of *P. formosana*.

A single clone introduced into cultivation by Messrs Charlesworth in 1912 was, unfortunately, mistakenly identified by Prain (1914) as *P. pogonioides* and has been grown under that name until the present day. However, *P. pogonioides* is certainly a synonym of *P. bulbocodioides* and the correct name for this clone is, in our opinion, *P. speciosa*. This clone (as *P. pogonioides*) was examined cytologically by Hunt & Vosa (1971) and was found to be a tetraploid. They also examined its

DNA content at the mitotic metaphase and found it to be similar to that of the tetraploid clone of *P. limprichtii*, but quite different from those of any of the diploid clones of *P. formosana* and *P. limprichtii* examined. In fact, all the tetraploids examined contained far less DNA than might have been expected from the simple doubling of the diploid chromosome number.

Polyploidy has undoubtedly played a large part in the evolution of this complex, but more material of this species must be studied before definite conclusions can be drawn.

We have now seen several clones of *P. speciosa* in addition to the one that has been in cultivation for many years. It is apparent that it is a variable species particularly in flower size, the intensity of its coloration, and the lip spotting and callus colour. The lip shape is however constant, with the red lip-spotting confluent, or almost so, in a band across the apical part of the mid-lobe. The callus-lamellae, of an even height and markedly sinuous, can be either yellow or white.

At the Royal Botanic Gardens, Kew there is a very small-flowered variant with deep purple flowers, recently collected by John Simmons and Hans Fliegner, in Kweichow Province, China.

Pleione speciosa is an easy species to cultivate and flower. Plants can produce two flowers per stem and also sometimes two flowering stems on a larger pseudobulb.

Pleione speciosa Ames & Schltr. in Feddes Repert. Beih. 4: 61 (1919); Cribb et al. in Curtis's Bot. Mag. 184: 140, t.868 (1983). Type: China; Hubei, *Wilson* 1761 (isotype K!).

Description. *A terrestrial herb. Pseudobulbs* conical, up to 30 mm long, 15 mm in diameter, rough-surfaced, 1-leafed at the apex. *Leaf* elliptic-lanceolate, up to 15 cm long or more, 2.5 cm wide or more, acute, only half developed at anthesis. *Inflorescence* up to 22 cm long, 1(–2)-flowered, the scape up to 14 cm long. *Bract* linear-lanceolate, 25–31 mm long, acute. *Flower* bright rose-purple with yellow lamellae on the lip. *Dorsal sepal* narrowly elliptic, 45–62(–70) mm long, 6–12 mm wide, acute. *Lateral sepals* obliquely narrowly elliptic, 40–55(–70) mm long, 7–13 mm wide, acute. *Petals* falcate, oblanceolate, 42–64 mm long, 5–10 mm wide, acute. *Lip* subrhombic to obovate, very obscurely 3-lobed, 42–55 mm long, 35–42 mm wide, with denticulate apical margins; callus consisting of 2 or 4 denticulate keels. *Column* 35–45 mm long. 2n = 40. plate 11; figure 8C, p. 69.

Distribution. China (Hubei, Sichuan, Kweichow & possibly Yunnan); altitude 1750–2250 m. map 4, p. 68.

LITTLE-KNOWN AND DOUBTFUL SPECIES

PLEIONE SAXICOLA

The description of this new species was received after the submission of the typescript of this book to the publisher. We have seen no authentic material of *P. saxicola* and the description given below is taken from the type description and a drawing of the type specimen which accompanies it.

Chen has compared *P. saxicola* to the well-known autumn-flowering species *P. praecox*, distinguishing it on its leaf-sheaths which lack any raised warts on their surface and on its lip-callus which comprises three entire lamellae rather than three to five papillate lines. No mention is made of either the flowering season or the exact locality of *P. saxicola*. It seems certain that Chen is mistaken in considering that this species is an ally of *P. praecox*. Apart from the characters he has given, he states categorically that the new species has one-leafed pseudobulbs. Both of the autumn-flowering species have two-leaved pseudobulbs. Examination of the published description and illustration suggests strongly that the true affinities of *P. saxicola* lie with *P. yunnanensis* and the *P. bulbocodioides* complex, all of which have one-leafed pseudobulbs and smooth sheaths on the new shoots. Altough we cannot be dogmatic, with the evidence to hand we consider that *P. saxicola* is very close to, if not conspecific with, the variable *P. bulbocodioides*. We have seen clones of the latter in which there are only three callus-lamellae, the central one running along the mid-vein of the lip. Chen states that the callus-lamellae are crispate, again a possible description of those of *P. bulbocodioides*. Until preserved or living material of this species can be studied, however, these ideas on relationship are merely speculative.

Pleione saxicola T. Tang & Wang ex S.C. Chen in Acta Phytotax. Sin. 25(6): 473 (1987). Type: China, Yunnan, without precise locality, 2400–2500 m, *K.M. Fang* 7914 (holotype PE).

DESCRIPTION. *A terrestrial herb. Pseudobulbs* ± turbinate or ovoid, 7–11 mm tall, 10–20 mm in diameter, with a persistent short annulus at the apex, 1-leafed. *Leaf* suboblong-lanceolate or oblanceolate, acute or obtuse, 10–18 cm long, 1.7–3.1 cm wide; petiole 3–7 cm long. *Inflorescence c.* 10 cm tall, 1-flowered. *Bract* oblanceolate, acute, 20–30 mm long, up to 10 mm wide, longer than ovary. *Flower* large, rose-coloured, *c.* 10 cm across (when flattened); pedicel and ovary 1.4–1.7 cm long. *Dorsal sepal* oblanceolate, acute, *c.* 63 mm long and 8 mm wide. *Lateral sepals* similar but oblique. *Petals* oblanceolate, narrower and slightly shorter than the sepals. *Lip* broadly elliptic in outline, 3-lobed towards the apex, *c.* 55 mm long, 22–25 mm wide, cuneate and unguiculate at the base; side-lobes broadly ovate, 4–5 mm wide, with irregularly crispate-crenulate margins; mid-lobe semi-orbicular, subrounded at the apex, 8 mm long, 15 mm wide, with an irregularly crispate-crenulate margin; callus consisting of 3 parellel, entire, crispate lamellae, 16 mm long, *c.* 1 mm high, running from the middle of the disc to the middle of the mid-lobe. *Column* slender, 40–46 mm long, dilated and narrowly winged towards the apex.

DISTRIBUTION. SW China (Yunnan); altitude 2400–2500 m.

Pleione chunii Tso in Sunyatsenia 1(2–3): 149 (1933). Type: China; Guangdong, Lockchong, *Chun* 43047 (not seen). MAP 4, p. 68.

We have not seen any authenticated material of *P. chunii* and thus its identity remains something of a mystery. Hunt & Vosa (1971) considered it to be conspecific with *P. yunnanensis*, but this is undoubtedly incorrect for *P. chunii* differs markedly in the shape and size of the sepals and petals and in having six fimbriate, rather than five entire, lamellae on the lip. From the original description it would appear to be most closely allied to *P. bulbocodioides* but it is possibly distinct in having mottled pseudobulbs, broader sepals and petals and six, rather than four or five, lamellae on the lip.

Pleione amoena Schltr. in Feddes Repert. Beih. 4: 185 (1919). Type: China; Sichuan, Nan Ch'uan, *von Rosthorn* 2131 (not seen).

Pogonia pleionoides Kraenzl. in Engl., Bot. Jahrb. 29: 267 (1901). Type: as for *Pleione amoena*.

These names are based on the same type material.

We have not seen any material of this species, but from the description of the type Schlechter was undoubtedly correct in assigning it to the genus *Pleione*. It may well be that the type specimen was destroyed in the Berlin Herbarium during the Second World War. Judging from Kraenzlin's description this plant would seem to be closest to *P. speciosa*, particularly in its large flower size, flower colour, lip shape and in the denticulate lamellae. The possibility remains that it may be conspecific with *P. speciosa*, in which case the earliest specific epithet for that taxon would be 'pleionoides'. However, until authenticated material has come to light such a move would be premature.

ARTIFICIAL HYBRIDS

Although the first hybrid, *Pleione* Versailles, was registered by Morel as long ago as 1966, it is only in the past ten years that the genus has been used at all extensively in hybridization programmes. So far over forty grexes have been registered involving, directly or indirectly, eight species. With a similar number of species either not in cultivation, or in cultivation but not yet used in crossing programmes, it is obvious that novelties will continue to be introduced over the next few years. A series of articles by Butterfield & Bailes (1986) presents the only up-to-date account of modern breeding.

Ambitious attempts have already been made, so far without success, to cross *Pleione* with species in allied genera such as *Coelogyne, Dendrochilum* and *Bletilla.* If eventually successful then the range of colour, form and climatic tolerance will be yet further extended.

The following is a list of some of the more popular and showy hybrid grexes currently available from specialist nurseries. It covers all those hybrids registered up to the present time, but as new grexes are being registered almost every month it may not be completely up to date.

Pleione Alishan (*P. formosana* × *P.* Versailles). First flowered during 1979. Pseudobulbs large, pyriform, dark olive-brown to purplish black. Flowers 2 per stalk, varying from pale to dark pink with very variable lip-marking; some cultivars, like that illustrated, which had *P.* Versailles 'Bucklebury' as one parent, have pronounced white tips to the sepals and petals. 2n = 60. Among the best cultivars are 'Martin' with pronounced white tips to the sepals and petals and an almost white lip; 'Sparrowhawk' with dark mauve-pink reflexed sepals and brownish red spotting on the lip; 'Goldfinch' with orange-yellow lip-spotting; 'Merlin' which is a vigorous grower and produces large flowers on a long stalk, with ice-pink sepals and petals fading to white at the tips, and a contrasting white lip boldly marked with red, PLATE 12.

Pleione Asama (*P. speciosa* × *P.* Vesuvius). PLATE 18E. First flowered in 1983. Most cultivars have the flower shape of *P. speciosa*, however a few take after *P.* Vesuvius in shape but with much richer colouring.

The colour varies from pale lavender to violet-purple, the lip being marked with red or rusty brown and with yellow lamellae.

Pleione Barcena (*P. formosana* × *P. praecox*). PLATE 17A. Pseudobulbs flattened and wrinkled, green with purple shading. Flowers produced in January, almost unicolored, mid lavender-pink, the lip with a paler, waved and frilled margin, stained orange-brown and spotted with purple within.

This fascinating hybrid is between a spring- and an autumn-flowering species. Like other such hybrids, the plants appear to go dormant for a short while after the leaves have dropped and the flowers appear before any roots develop. As a result, care should be taken not to overwater at this stage, the aim being to keep the compost just moist. A temperature of 5°C or slightly higher is ideal during the winter months.

Pleione Brigadoon (*P. speciosa* × *P.* × *confusa*). PLATE 12. Various forms of this grex exist. Those with *P. speciosa* as the female parent have purplish, pyriform pseudobulbs. The flowers, often 2 per stem, are pale mauve shading to violet-purple, with the lip marked with red and yellow, or sometimes with a solid red base. 2n = 40. Select cultivars of this cross include: 'Stonechat' with dark petals and a dark rose lip suffused with orange; 'Woodcock' with slightly paler petals but an orange lip.

Pleione Cotopaxi (*P. limprichtii* × *P.* Vesuvius). Pseudobulbs pointed, dark olive-green. Flowers pale lilac, faintly shaded with puce, the lip well frilled and heavily spotted with crimson and with yellow lamellae. 'Treecreeper' is a particularly fine cultivar with a very frilled lip spotted with red.

Pleione Danan (*P. limprichtii* × *P. humilis*). Pseudobulbs tall, olive-green, resembling those of *P. humilis*. Flowers amethyst-violet, the lip white with violet shading and often heavily spotted with dark red. Some clones also have some yellow on the lip. 'Chaffinch' is particularly fine, with a dark flower with an extra large and heavily spotted lip.

Pleione Eiger (*P. formosana* × *P. humilis*). PLATES 14, 17B. Pseudobulbs large, somewhat pyriform, green. Flowers 2 per stem, produced during January and February, mostly white shaded with pink, the lip with red, or red and yellow markings. 2n = 40. Some cultivars have a white lip with yellow markings, whilst others have the petals

A

B

C

D

E

F

Plate 15 **A**, *Pleione praecox*, growing on a roadside bank south of Kathmandu, Nepal (photo. C. Bailes); **B**, *Pleione humilis* in the Gurkha Himal, Nepal (photo. C. Grey-Wilson); **C**, *Pleione hookeriana*. A rare white clone from Nepal (photo. I. Butterfield); **D**, *Pleione maculata*, in cultivation (photo. H. Pfennig); **E**, *Pleione scopulorum*, a lilac-flowered clone from south-west China (photo. H. Pfennig); **F**, *Pleione scopulorum*, a darker purple clone from south-west China (photo. H. Pinkepank).

A

B

C

D

E

F

Plate 16

A, *Pleione forrestii*, a plant shown at the 12th World Orchid Conference in Tokyo (photo. P. Cribb); **B**, *Pleione* × *confusa*, cultivated and photographed by I. Butterfield; **C**, *Pleione bulbocodioides* growing in the Baishui valley, Lijiang, north-west Yunnan (photo. C. Grey-Wilson); **D**, **E**, *Pleione bulbocodioides* growing on the slopes of Yulongxue Shan, north-west Yunnan (photo. P. Cribb); **F**, *Pleione coronaria* cultivated and photographed by I. Butterfield.

A

B

C

D

E

F

Plate 17

A, *Pleione* Barcena cv. (photo. I. Butterfield); **B**, *Pleione* Eiger cv. (photo. P. Cribb); **C**, *Pleione* Irazu 'Mallard' (photo. I. Butterfield); **D**, *Pleione* Matupi 'Golden Oriole' (photo. I. Butterfield); **E**, *Pleione* Shantung 'Fieldfare' (photo. I. Butterfield); **F**, *Pleione* Paricutin 'Chiffchaff' (photo. I. Butterfield).

Plate 18 **A**, *Pleione* Rakata 'Shot Silk' (photo. P. Cribb); **B**, *Pleione* Rakata 'Nuthatch' (photo. I. Butterfield); **C**, *Pleione* Erebus cv. (photo. P. Cribb); **D**, *Pleione* Tolima 'Moorhen' (photo. I. Butterfield); **E**, *Pleione* Asama cv. (photo. I. Butterfield); **F**, *Pleione* Versailles 'Bucklebury' (photo. M. Svandelik, RBG, Kew).

A

B

C

D

E

F

Plate 19

A, *Pleione formosana* 'Snow White' (photo. P. Cribb); **B**, *Pleione praecox* 'Everest' (photo. M. Svandelik, RBG, Kew); **C**, Selected plants from a hybrid swarm of *Pleione bulbocodioides* × *P. yunnanensis* from Cang Shan, west Yunnan (photo. I. Butterfield); **D**, *Pleione* Shantung cv., cult. Kew (photo. P. Cribb); **E**, *Pleione* Versailles 'Bucklebury' (photo. P. Cribb); **F**, *Pleione* El Pico 'Pheasant' (photo. I. Butterfield).

A

B

Plate 20 **A**, Habitat of *Pleione bulbocodioides* in north-west Yunnan, Yulongxue Shan, Baishui (photo. P. Cribb); **B**, Habitat of *Pleione humilis*, Dharchy Forest, Gurkha Himal, Nepal (photo. C. Grey-Wilson).

flecked with red. 'Lesley Frank' has large creamy flowers with yellow and chestnut-red spotting on the lip.

Plants flower before root initiation takes place. *Pleione* Ruth, though registered as a distinct grex, must be included here as its parentage is *P. pricei* × *P. humilis*. *Pleione pricei* is considered to be a synonym of *P. formosana*.

Pleione El Pico (*P.* Versailles × *P. bulbocodioides* (6 *x*)). Pseudobulbs flattish, dark green or rarely purplish. Flowers appearing rather late in the season, pale mauve-pink to dark rosy purple, some forms with very pale lip-markings, others blotched with rich red. 2n = ±100. The best cultivars, produced when *P.* Versailles 'Bucklebury' is used as a parent, include: 'Goldcrest' with a small dark flower on a long stem and a dark red-blotched lip with a yellow callus; 'Kestrel' with a paler flower; 'Pheasant' with a large dark, heavily red-spotted flower, PLATE 19F; 'Starling' with similar flowers to 'Pheasant' but flowering later.

Pleione Erebus (*P.* Versailles × *P.* Vesuvius). PLATE 18C. First flowered in 1982. Pseudobulbs variable from squat to pyriform, green to purplish black. Flowers pale violet-purple, the lip white spotted with dark red and with yellow lamellae. This cross has produced several fine cultivars, such as 'Redshank' with an orange-yellow lip and 'Willow Warbler' with very pale reflexed petals and a pale creamy yellow, brown-spotted lip.

Pleione Etna (*P. speciosa* × *P. limprichtii*). Pseudobulbs pyriform, dark to blackish green. Flowers 2 per stem, slightly paler than *P. speciosa* but with the lip-markings of *P. limprichtii*. 2n = ±80. The large dark-flowered cultivar 'Bullfinch' is particularly fine.

Pleione Fuego (*P. formosana* × *P. bulbocodioides* (6 *x*)). Pseudobulbs small, dark greenish. Flowers prolifically produced. 2n = ±80. Those with *P. formosana* 'Oriental Splendour' as one parent result in rather smaller, long-lasting flowers of a rich purplish pink, the lip well frilled and marked with blood-red spots and with white lamellae within. Those with *P. formosana* 'Clare' as a parent have large, flattened green pseudobulbs and large pale mauve-pink flowers with a spotted lip. 'Wren' is one of the most attractive cultivars which has a small dark flower, with a prominent pale yellow callus and the flowers borne on a long stem.

Pleione Fujiyama (*P*. El Pico × *P*. Shantung). First flowered in 1986. Colour varies from pale petunia-purple to dark mauve-pink with the lip mostly yellow, heavily marked with red.

Pleione Fu Manchu (*P. speciosa* × *P*. Eiger). Made by Hazelton in 1984, but not seen by us yet.

Pleione Hekla (*P. speciosa* × *P. humilis*). PLATE 14. First flowered in 1982. Pseudobulbs tall, pyriform, dark green to purplish. Flowers elegant, petunia-purple with a frilled lip blotched with dark red and with yellow lamellae. The cultivar 'Partridge' is larger and darker-flowered than most.

Pleione Helgafell (*P*. Eiger × *P. yunnanensis*). First flowered in 1986. A very pale lilac flower with faint purple spotting on the sepals, and reddish brown markings and yellow lamellae on the lip.

Pleione Irazu (*P*. Etna × *P*. Shantung). Pseudobulbs smallish, dark green. Flowers mallow-purple with an almost white lip with yellow lamellae and heavy brownish red blotching. The cultivar 'Mallard' has dark petals and a yellow lip with heavy red spotting, PLATE 17C.

Pleione Jorullo (*P. limprichtii* × *P. bulbocodioides* (6 *x*)). Pseudobulbs variable, dark green to blackish purple. Flowers small, pale violet-purple, rather like *P. limprichtii* in shape, the lip with large crimson spots and blotches.

Pleione Katla (*P. limprichtii* × *P*. Versailles). Pseudobulbs like those of *P. limprichtii* but more robust. Flowers also similar, though larger. $2n = \pm 80$. The heavy dark spotting on the lip of the cultivar 'Mistle Thrush' is particularly noteworthy.

Pleione Kilauea (*P. formosana* × *P*. Eiger). *Pleione formosana* 'Oriental Splendour' was used to produce this cross which first flowered in 1986. The sepals are various shades of pinkish mauve with the lip paler, almost white, marked with red or purplish brown or occasionally with yellow. The cultivar 'Hoopoe' has dark mauve-pink sepals and petals and solid dark red lip-markings.

Pleione Lascar (*P*. Fuego × *P. limprichtii*). First flowered in 1986. The influence of *P. limprichtii* in the progeny is considerable. Some cultivars have a very open lip with large red spots. In others the lip-markings are crimson, orange-brown or purple.

Pleione Lassen Peak (*P. praecox* × *P*. × *lagenaria*). First flowered in 1984. Sepals and petals amethyst-violet, the lip almost white with purple-violet marginal blotches. The flowers are fragrant and darken with age.

Pleione Lipari (*P. bulbocodioides* × *P.* Vesuvius). Registered by Butterfield in 1982. Most clones have rose-purple flowers and a red-spotted lip.

Pleione Matupi (*P. limprichtii* × *P.* Shantung). First flowered in 1983. Flowers are all shades of pinkish mauve. Some but not all of the lips are yellow and are marked with orange-red or reddish brown. 'Golden Oriole' has particularly fine, dark yellow, fragrant flowers with a mauve sheen, PLATE 17D. The flowers of 'Corncrake' are pale old rose with a yellow lip mottled with brown.

Pleione Nero Wolfe (*P. praecox* × *P. bulbocodioides*). Raised and registered by H. Pinkepank in 1985. We have not yet seen any plants.

Pleione Novarupta (*P.* Versailles × *P.* Soufrière). First flowered in 1986. Clones vary in sepal and petal colour from pale amethyst to plum-purple. The lip also varies from pale to deep yellow, sometimes flushed with pink on the outside. The red lip-markings can be very heavy or sparse. 'Goshawk' is an especially fine clone with purplish sepals and petals and a yellow lip.

Pleione Paricutin (*P.* Tongariro × *P.* × *confusa*). First flowered in 1986. Most clones are purplish violet with ruby-red markings and a yellow callus on the lip. A few have a yellow or even orangish lip. The cultivar 'Chiffchaff' has pale salmony mauve flowers with a yellowish lip, PLATE 17F.

Pleione Piton (*P. formosana* × *P. yunnanensis*). First flowered in 1986. The long-lasting flowers are almost uniformly pale lavender with purplish red or violet markings on the lip.

Pleione Rakata (*P. speciosa* × *P.* Shantung). Pseudobulbs tallish, purple-black. Flowers mauve-pink, the lip with orange spotting and blotching. Several of the finest hybrid cultivars belong here; 'Blackbird' has a very dark flower; 'Nuthatch' has a pale orange lip, PLATE 18B; 'Redwing' has a large dark flower with a dark lip spotted with dark red; 'Shot Silk' is outstanding with deep salmon-orange flowers with magenta overtones, the lip slightly paler with orange and red spots, PLATE 18A.

Pleione Sajama (*P. limprichtii* × *P. hookeriana*). First flowered in 1986. The flower shape and colour show a strong influence from

P. limprichtii but the lip is long and open with quite large red-brown and purplish red spots within.

Pleione Sangay (*P. limprichtii* pink form × *P. praecox*). Pseudobulbs somewhat resembling *P. praecox*, dark green to purplish green. Flowers pale lavender-pink, the lip almost white inside and with brown and pale purple spots. The leaves turn an attractive purplish pink in the autumn.

Pleione Shantung (*P. formosana* × *P.* × *confusa*). PLATES 13, 19D. Pseudobulbs very large, green to dark purple. Flowers dark yellow, or white flushed with pink, or pink, the lip marked with red, often fragrant. 2n = ± 60, ± 66, ± 67. The best-known and most vigorous of all hybrids. 'Muriel Harberd' is the largest apricot-flowered clone. The cultivar 'Fieldfare' has large pale yellow flowers with dark red-brown lip-markings, PLATE 17E. 'Golden Plover' has pale yellow sepals and petals, flushed with mauve-pink, and a dark yellow lip with red spots; 'Silver Pheasant' has an almost white flower with red lip-markings.

Pleione Sorea (*P. hookeriana* × *P. bulbocodioides*). First flowered in 1983. Produces relatively small but darkly coloured flowers with heavily frilled lips boldly spotted with red; the callus-lamellae are prominent.

Pleione Soufrière (*P.* Versailles × *P.* × *confusa*). PLATE 13. First flowered in 1981. Pseudobulbs large, flattened on the flower-bud side, dark green. Flowers 2 per stalk, variable, pale rose-purple to amethyst-violet, the lip pale yellow, heavily marked with dark red.

Pleione Stromboli (*P. speciosa* × *P. bulbocodioides* (6 *x*)). First flowered in 1979. Pseudobulbs large, flattish, dark green to purplish-black. Flowers on tall stems, mostly dark reddish pink, the lip marked with purplish red, often scented. 2n = ± 100. 'Fireball' is one of the most intensely coloured of all *Pleione* cultivars, with glowing magenta flowers whose lip has rich red markings in the top half and between the callus-lamellae, PLATE 12.

Pleione Surtsey (*P.* Shantung × *P.* Versailles). First flowered in 1983. Clones vary in colour from lavender-pink to pinkish mauve and have broad open lips of a paler shade but with brownish spotting. The cultivar 'Hawfinch' has pale salmon sepals and petals that are shaded with mauve and a pale straw-yellow lip with soft brown spotting.

Pleione Tameraine (*P.* Alishan × *P.* Tolima). Raised by Hazelton and registered in 1984. We have not yet seen any plants.

Pleione Tarawera (*P.* Versailles 'Bucklebury' × *P. praecox*). Pseudobulbs purplish or dark green. Flowers pale to dark mauvish pink, the lip varying from whitish to dark mauve-pink, always heavily blotched with crimson-red and a little yellow.

Pleione Tolima (*P. speciosa* × *P. formosana*). First flowered in 1979. Pseudobulbs pyriform, purplish black. Flowers 2 per stem, very variable, mauve-pink to very dark rose-purple, the lip unspotted to heavily spotted with red and with yellow or red lamellae, sometimes scented. 2n = 60. This cross had produced several attractive cultivars: 'Avocet' with a soft pinkish mauve flower with a very frilly lip; 'Nightingale' has a paler fragrant flower with a few red spots on the lip; 'Moorhen' has a dark flower with a prominent yellow callus, PLATE 18D; 'Waxwing' has a large dark flower with the area along the callus being orange-yellow.

Pleione Tongariro (*P.* Versailles × *P. speciosa*). Flowers 2 per stem, long-lasting, various shades of imperial purple, the lip marked with red and yellow and with yellow lamellae. 2n = ±80. The cultivar 'Jackdaw' has a dark flower with dark red spotting on the lip and an orange-yellow callus.

Pleione Tsingtau (*P. humilis* × *P. praecox*). Raised and registered by H. Pinkepank in 1985, but we have not yet seen any plants.

Pleione Versailles (*P. formosana* × *P. limprichtii*). The first hybrid *Pleione* ever raised, it flowered about 1962–3. Flowers 2 per stem, variable in colour, very pale mauvish pink with slightly darker lavender lip-markings to darker rose-pink with deep blood-red lip-markings; some forms have yellow or orange-brown markings on the lip. 2n = 40, 60. 'Bucklcbury' is a particularly fine cultivar with large, dark green pseudobulbs, each with 2 or 3 flowering stems bearing 2 flowers per stem; it was raised by Professor Morel, PLATES 18F, 19E. Mr Brian Williams, then the grower for Lady Sainsbury at Bucklebury, Berkshire received a First Class Certificate for this cultivar from the Royal Horticultural Society. It has been widely used as a parent of more recent hybrids.

Other worthwhile cultivars include 'Heron' with a pale flower on a long stem; 'Linnet' with dark sepals and petals and an almost white lip marked with red-brown; 'Muriel Turner' with pinkish mauve flowers, spotted heavily on the lip; 'Puffin' has a large flower in which the lip is spotted with red-brown; 'Swallow' has a small pale lavender-pink flower with darker spotting on the lip, PLATE 14.

Pleione Vesuvius (*P. bulbocodioides* (6 *x*) × *P.* × *confusa*). First flowered in 1978. Pseudobulbs large and flat with prominent flower-buds, dark greenish black. Flowers variable in colour from pale lavender or pink to dark mauve, the lip marked with red or red and yellow, occasionally orange-brown with red spots, sometimes scented. 2n = ±78 to ±84.

When *P.* × *confusa* is used as the female parent the offspring are slower growing and the flowers are pale lavender-pink with subdued lip-markings.

Vesuvius has produced several notable cultivars including 'Grey Wagtail' with a pale lavender-pink flower with dark red markings on the lip; 'Leopard' with a dark flower, the red-orange lip spotted with red; 'Phoenix' with a dark flower with an orange-brown lip spotted with red, PLATE 13; 'Redstart' with a dark flower with an even darker lip with a yellow callus; 'Snipe' with a pale flower with a yellow lip suffused with pink and spotted with red-brown.

Pleione Wunzen (*P.* Erebus × *P. yunnanensis*). First flowered in 1986. Produced small pale mauve-pink long-lasting flowers with a paler lip heavily spotted with reddish violet.

Pleione Yu Shan (*P. pricei* × *P. speciosa*). Registered by Hazelton in 1983, but synonymous with *P.* Tolima.

ADDENDUM

After the typescript of this book had been submitted to the publisher, we received six *Pleione* plants from Dr H. Pfennig of Herford in West Germany. Two of these were in full flower and the remainder in advanced bud. Following careful examination of the flowering specimens and comparison of them with the herbarium material preserved at Kew, we are convinced that they represent a new species. Because the plants arrived too late for this beautiful species to be included in the main text of this book we are describing it below.

Pleione aurita Cribb et Pfennig, **sp.nov***., affinis *P. hookerianae* (Lindl.) B.S. Williams sed pseudobulbis majoribus, inflorescentia longiore, flore majore distincte colorato, labello emaculato sed in centro lamellae aurantiaci-flavo et 4–5 lineis parallelis pilis longis ornatis satis distinguenda. Typus: China, Yunnan, fl. in cult. 22 March 1988, *Pfennig* s.n. (holotypus K!).

DESCRIPTION. *A terrestrial herb* up to 16 cm tall. *Pseudobulbs* conical, angular in cross-section, 20–45 mm long, 10–20 mm in diameter, green or pale green. *Inflorescence* erect, 1-flowered. *Peduncle* 50–70 mm long. *Bract* narrowly elliptic-oblanceolate, acute or subacute, cucullate, 25 mm long, 10–15 mm wide, pale pink with darker veins. *Flowers* showy, pale pink, rose pink or purple, paler towards the base of the floral segments; lip with a central longitudinal yellow or orange-yellow stripe dilated at the apex; callus-hairs orange or yellow; pedicel and ovary arcuate, 6-ribbed, 1.6–2 cm long, green. *Dorsal sepal* hooded over column, narrowly elliptic or oblong-elliptic, subacute, 42–50 mm long, 11–15 mm wide. *Lateral sepals* porrect, not spreading widely, obliquely elliptic, subacute, 40–43 mm long, 14–16 mm wide. *Petals* strongly reflexed, oblanceolate or spathulate, obtuse or rounded, 40–43 mm long, 14–19 mm wide. *Lip* deeply cucullate, obscurely 3-lobed towards the apex, broadly flabellate when flattened, emarginate, 39–40 mm long, 50–60 mm wide, undulate and irregularly erose on the apical margin; callus of 4 or 5 lines of long hairs from base to 5 mm from apex of the lip. *Column* clavate, dilated or narrowly winged at the apex, 25–27 mm long, notched on each side at the apex. FIGURE 9.

* As published in Die Orchidee 39(4) – (1988).

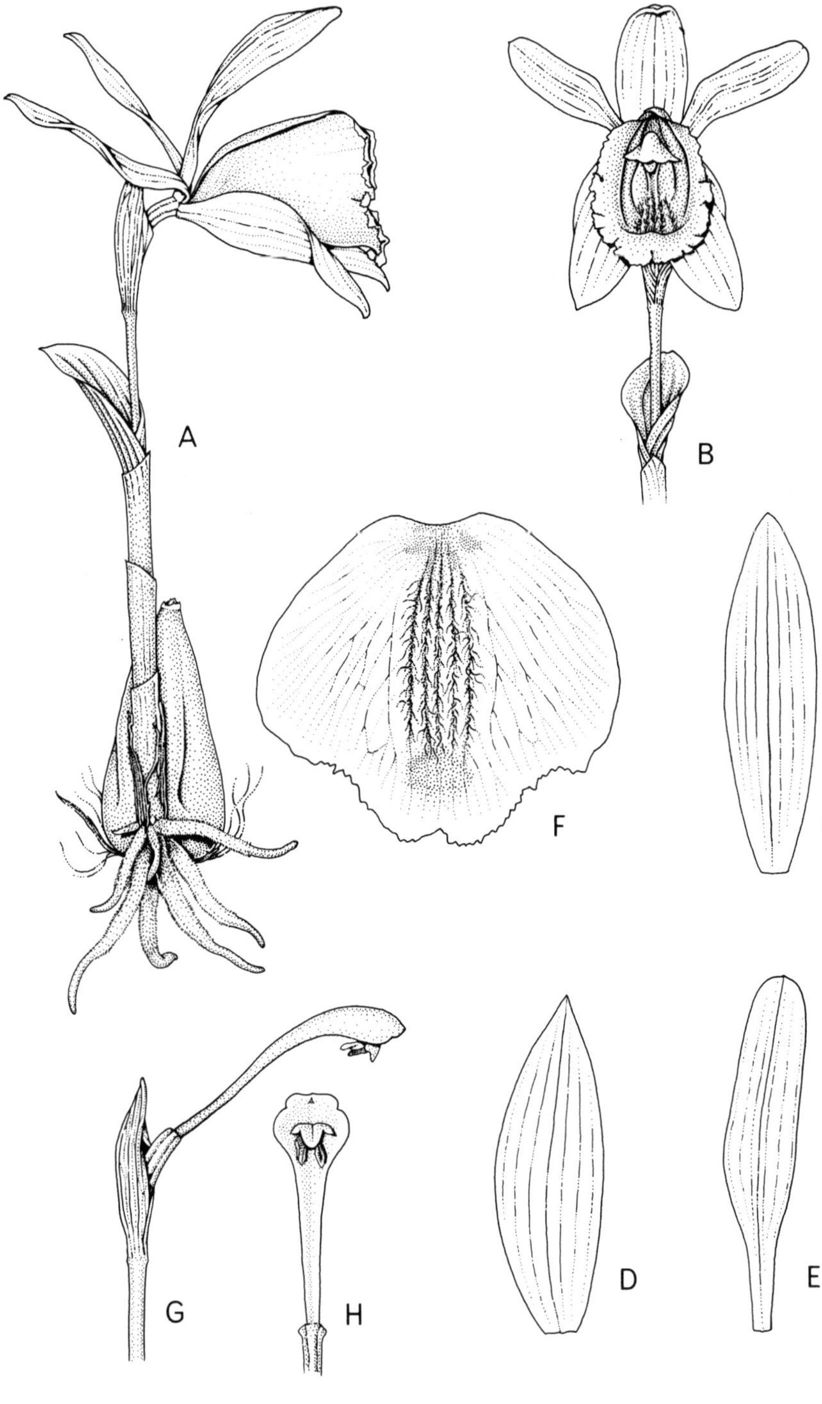
A
B
F
C
G
H
D
E

DISTRIBUTION. China: Yunnan.

Pleione aurita is a spectacular new species which is apparently most closely allied to the widespread *P. hookeriana*. Both are characterized by their one-leafed pseudobulbs and flowers in which the lip is broader than long and bears a callus of several lines of long hairs reaching from the base almost to the apex. It differs, however, in its very much larger pseudobulbs and overall habit and in its distinctively coloured flowers. The flowers vary in shade from pale pink to deep bright purple and the lip has a central longitudinal yellow or orange-yellow stripe which is dilated at the apex and terminates a few millimetres from the apex. The lip is also distinctive in having an undulate margin and a notched apex while the callus comprises four or five rows of long, yellow hairs on the central veins.

Nothing is known of the habitat of this attractive species but the material sent to Kew is said to have come from Yunnan. It was received by the grower as *P. forrestii* which suggests that it might have originated in western Yunnan.

The specific epithet has been given for the distinctive attitude of the petals which resemble the ears of a rabbit or hare in their shape and posture.

Figure 9. (opposite) *Pleione aurita*. A, habit, ×2/3; B, flower, ×2/3; C, dorsal sepal, ×1; D, lateral sepal, ×1; E, petal, ×1; F, lip flattened out, ×1; G, column, ovary, pedicel and bract, ×1; H, column, ventral view, ×1. All drawn from the type by Timothy Galloway.

BIBLIOGRAPHY

Arditti, J. (1982). Orchid seed germination and seedling culture—a manual. In Arditti, J. (ed.), *Orchid Biology: Reviews and Perspectives* II: Appendix 243–370. Ithaca & London.

Butterfield, I. & Bailes, C. (1986). Recent advances in Pleione breeding. *Orchid Rev.* 94: 63–7, 91–4, 134–6, 237–9.

Cooper, F.J.C. (1965). Pleione humilis 'Frank Kingdon Ward'. *Bull. Alpine Gard. Soc.* 33: 251–2.

Cribb, P.J. & Page, F. (In prep.). Seed morphology of Pleione species.

Cribb, P.J., Tang, C.Z. & Butterfield, I. (1983). The genus Pleione. *Curtis's Bot. Mag.* 184: 93–147.

Day, J. (1866). Pleione maculata and P. lagenaria. *Scrapbook* 11: tt. 29, 30. Unpubl. Herb. Kew.

Harberd, D. (1980). Breeding new cultivars in the genus Pleione. In Brickell, C.D., Cutler, D.F. & Gregory, M. (eds.), *Petaloid Monocotyledons* 171–81. London & New York.

Hooker, J.D. (1890). Coelogyne Lindl. In *Fl. Brit. India* 5: 828–43.

Hunt, P.F. (1962). Pleione limprichtii. *Curtis's Bot. Mag.* 174: t. 397.

——— & Vosa, C.G. (1971). A cytological and taxonomic study of the genus Pleione D. Don (Orchidaceae). *Kew Bull.* 25: 423–32.

Johnson, M. (1986). The hardy Chinese orchid Pleione limprichtii. *Kew Mag.* 3: 73–8.

Kretz, R. (1987). Destination Rupina La. *Bull. Alpine Gard. Soc.* 55: 297–304.

Lancaster, R. (1982). Five orchids of Yunnan. *The Garden* 107: 425–30.

Lindley, J. (1830). Coelogyne. *Gen. Sp. Orchid. Pl.* 38–44.

——— (1854). *Fol. Orchid.* 5. *Coelogyne*: 1–18.

Morel, G. (1971). Les "Pleiones". Culture—Multiplication. *Orchidophile* 6: 88–94.

Pfitzer, E. & Kraenzlin, F. (1907). Pleione D. Don. In Engler, A., *Pflanzenr.* (IV.50): Coelogyninae 119–29.

Prain, D. (1914). Pleione pogonioides. *Curtis's Bot. Mag.* 140: t.8588.

Reichenbach, H.G. (1864). Coelogyne. In Walpers, *Ann. Bot.* 6: 222–39.

Rolfe, R.A. (1903). The genus Pleione. *Orchid Rev.* 11: 289–92.

Schlechter, R. (1914). Die Gattung Pleione und ihre Arten. *Orchis* 8: 72.

——— (1919). 56. Pleione D. Don. *Feddes Repert. Beih.* 4: 185–7.

——— (1927). 156. Pleione. *Die Orchideen* (ed. 3), 148–50.

Stergianou, R. (1987). *Breeding Studies in Pleione* (*Orchidaceae*). PhD Thesis, University of Leeds.

Thompson, P.A. (1977). *Orchids from seed.* London.
Williams, B. (1980). *Orchids for everyone.* London.
Wimber, D. & Cribb, P.J. (1981). A cytological study of the species and hybrids in the genus Pleione (Orchidaceae). *Plantsman* 3: 178–88.

INDEX TO SPECIFIC AND INFRASPECIFIC EPITHETS

Accepted names are given in roman type; synonyms are in *italic* type. Occasionally synonyms included in the index are not mentioned in the main text.